Design Heuristics for Emerging Technologies

*To my amazing kids, Vigo and Saka—your energy
and imagination are constant sources of inspiration.
Chewie, my faithful German shepherd, thank you for keeping
my feet warm and my heart full during those cold, early
mornings of writing. And to Stefan, whose love and
encouragement have been my anchor through this entire
process—I couldn't have done it without you.*

You are my greatest blessings.

Design Heuristics for Emerging Technologies

Artificial Intelligence, Data, and Human-Centered Futures: Considerations for the Rights of Women

Kem-Laurin Lubin

This is a work of nonfiction.

Design Heuristics for Emerging Technologies

First Edition: May 2025

Cover design by Cheyanne Rosier

Email Address: CheyRoseDesign@gmail.com

Printed in the United States of America

26 27 28 29 30 10 9 8 7 6 5 4 3 2 1

Mailing/Submissions:
Universal Write Publications, LLC
421 8th Avenue, Suite 86
New York, NY 10116

Website: UWPBooks.com

Print ISBN: 978-1-942774-44-0
eBook ISBN: 978-1-942774-45-7

Library of Congress Control Number: 2025901240

This book has been partially supported with a financial grant from SAGE Publishing.

Contents

Foreword

Lai-Tze Fan, PhD

In the rapidly evolving landscape of artificial intelligence (AI), we find ourselves at a critical juncture where technology's promises collide with its perils. Kem-Laurin Lubin's book, *Design Heuristics for Emerging Technologies*, arrives at this moment—when AI systems are being deployed with rapid speed across many social domains, often with inadequate consideration for their social implications.

What makes Lubin's work valuable is its examination of how AI systems reinforce existing power structures, with a focus on AI's impact on women and other marginalized groups. Her opening example of predictive algorithms targeting young girls in Argentina illustrates perfectly how technology marketed as progress can function as a sophisticated form of control and surveillance. This is not merely a technical problem—it is fundamentally a problem of power, values, and whose perspectives shape our digital future.

Drawing on over two decades of leadership in design research at organizations like BlackBerry, Autodesk, and Siemens, Lubin brings a rare combination of technological industry expertise and critical perspective from the arts and humanities. In this way, complicating existing discussions of AI ethics as a philosophical exercise, *Design Heuristics for Emerging Technologies* grounds its analysis in practical experience with digital transformation and the realities of how technologies are conceived, built, and deployed. This inside knowledge allows for a nuanced critique that goes beyond surface-level concerns to address the structural issues at the heart of technological design.

But what distinguishes Lubin's work from the growing library of works on AI ethics is its commitment to solutions. While many authors excel at diagnosing the problems of algorithmic bias, surveillance capitalism, and digital discrimination, Lubin takes the next step by offering concrete heuristics and design interventions. Her work refuses to accept technological determinism, instead insisting that because AI systems are designed, they can—and must—be *redesigned* with justice as their foundation.

Readers will find a thoughtful integration of feminist theory, technical analysis, and practical guidance. The book invites us to ask essential questions about AI systems: Who benefits? Who bears the costs? Whose knowledge and experience are privileged, and whose are erased? Through compelling case studies and clear frameworks, it demonstrates how to move from critique to creation, from identifying problems to implementing solutions.

For designers and technologists, *Design Heuristics for Emerging Technologies* will offer practical tools to build more equitable systems. For policymakers, it will provide initial frameworks to evaluate the societal impacts of AI. For activists and community organizers, it suggests strategic intervention points in technological systems.

Lubin's work positions itself within a broader movement that recognizes technology as a site of both oppression and potential liberation. By centering the experiences and knowledge of those most affected by technological systems, Lubin charts a path toward more just and equitable futures. She reminds us that the technical is always political, and that design choices embed values that shape society in profound ways.

As the reader encounters this work, perhaps it is most pertinent to remember that AI is not inevitable in its current forms. AI is designed, and we can design it differently, with women's rights, human dignity, and collective liberation in deeper consideration.

—Dr. Lai-Tze Fan, April 2025
Toronto, Canada

Preface

Lesley-Ann Noel, PhD

When I created the Critical Alphabet in 2018, I was angry! I had moved to Silicon Valley for a one-year fellowship. During this period, I was able to eavesdrop on, or occasionally participate in, conversations about designing new apps, technology, and programs. Why was I angry? I was frustrated that people believed discussions about positionality, race, poverty, marginalization, and more were unnecessary as they aimed to build their new technological futures. In some of these conversations, when I brought up a discussion point about how racial background, poverty, gender, and our intersectional identities could affect the products that we were brainstorming about, I'd often get shushed. I remember one collaborator even saying, "Hey, this is Silicon Valley, not a sociology class." Fortunately, times have changed somewhat, and conversations about intersectionality have become much more common in the design studio.

I studied industrial design (and a touch of graphic design) in the 1990s. At that time, we had little understanding of our roles in the design process; we thought we were neutral. We often researched materials as well as current and future design trends. Then, we followed a creative ideation process and, finally, presented our "findings." We did not consider how our worldviews, where we were born, or our genders, tastes, personal biases, past traumas, and more influenced the design process. Twenty years later, while taking a doctoral-level research methods elective at my university's College of Education, I learned the significance of understanding my positionality and how my intersectional identity shaped my worldview, ultimately affecting my design choices.

I've focused on encouraging designers to have deeper conversations and to have a critical awareness of the world around them, which Paulo Freire described as *conscientização*. This mission has driven my teaching and my research for several years. I've seen the design world and designers' interest in intersectionality evolve and grow over the last few years. Freire's theories about building a critical awareness of the world are tied to the concept of *agency*, since this awareness of the world also leads to the understanding that the world is not fixed, and we have the agency to change it.

Just as designers have had an incredible surge in the awareness that they are not neutral in the design process, people in the world of tech and AI must also develop a greater understanding that AI systems are not neutral but designed with the embedded values of the people who create them and the contexts in which they are developed. Therefore, the biases of the makers of technology eventually transfer to the outputs of that technology. One way to mitigate these dangers is to reflect on, analyze, and challenge the power structures and norms that surround the technology as it is created. This might help make visible the master–slave relationships between users and voice-controlled AI, or ensure that misogyny, racism, ableism, and harmful isms and structures from the human world are not transferred to the technology and AI realm.

Kem-Laurin Lubin draws on her many years of experience working in tech industries and blends it with her academic interests. She uses this experience to address broad societal challenges, championing ethical questions that her tech colleagues might ignore. Her work, like mine, is driven by Freirean ideals like promoting user agency and building a critical awareness of the technological world around us. In her book *Design Heuristics for Emerging Technologies: Artificial Intelligence, Data, and Human-Centered Futures,* Lubin aims to raise critical awareness of issues of power and ownership in the development and deployment of AI. She ensures readers reflect on who benefits from or is harmed by these systems and how design processes can center the needs and voices of those most affected by critical issues. While many people are critical of AI, she aims beyond mere critique to provide actionable strategies for designing AI and tech with a justice-centered approach. In the book, she offers case studies, frameworks, and solutions that empower designers, technologists, and activists, and she emphasizes the importance of feminist and community-centered approaches to AI.

In a world of intense political and social change and rising surveillance of rights, it is appropriate to reflect on the limits and capabilities of artificial intelligence and how we interact with these technologies. It is also relevant to remember our individual and collective agency and how we both make and shape the worlds that we live in. Artists, designers, and changemakers can play an active role in reshaping the technologies that we use for collective liberation, and Kem-Laurin Lubin shows us how we can do this.

—Dr. Lesley-Ann Noel, April 2025
Author of Design Social Change

Acknowledgments

I am deeply grateful to the many individuals whose support has been essential throughout the journey of writing this book. First and foremost, I extend my heartfelt thanks to my academic supervisors, Dr. Randy Harris and Dr. Lai-Tze Fan from the University of Waterloo, Canada. Your invaluable guidance, constructive feedback, and unwavering encouragement have been instrumental in shaping this work. I am also sincerely appreciative of my committee members, Dr. Marcel O'Gorman and Dr. Danielle Deveau, whose insightful comments and expertise guided the trajectory of this project, offering essential perspectives that have deepened and expanded my work. Additional gratitude goes to Dr. Lai-Tze Fan and Dr. Lesley-Ann Noel for their profound insights, generous support, and contributions to both the preface and foreword that significantly enhanced the clarity, depth, and accessibility of this book.

Importantly, I am deeply thankful to Dr. Ayo Sekai, whose insightful recognition of the broader implications of my work significantly expanded its impact and scholarly depth. Her encouragement and vision have inspired me immensely.

Special acknowledgment is extended to Samantha Blostein, director at Research for Change, and Dr. Catherine Burns of the University of Waterloo for their invaluable contributions, as well as their excitement about this project. Both Samantha and Catherine demonstrate exceptional passion and commitment to empowering women in their respective fields, championing women's achievements, and advocating for their advancement in both analog and digital spheres.

Lastly, I am also profoundly appreciative of my editors and book designer for their meticulous attention and dedication. Melinda Masson and Michelle Ponce, thank you for your thoroughness; Cheyanne Rosier—a striking book cover design is everything— thank you for your creativity.

To each of you, thank you for your invaluable contributions, unwavering support, and belief in my work. This journey would not have been possible without you.

Introduction

Why a supplement on **artificial intelligence (AI)** design considerations for the rights of women?

In 2017, Juan Manuel Urtubey, the conservative governor of Argentina's Salta province, made an announcement that stopped me in my tracks. But it would be two years later that it ultimately set me on a path of deeper inquiry into **data feminism** and female agency.

Standing before the public, Urtubey proudly declared a partnership with a Microsoft subsidiary to combat teenage pregnancy using AI. "With this technology," he proclaimed, "you can predict five or six years in advance—with the first and last name and address—which girl, future teenager, has an 86% predestination of having a teenage pregnancy" (Calacci, 2022).

AI forecasting the reproductive destiny of little girls? By name? What could possibly go wrong?

As it turns out, feminists can tell us exactly what could—and did—go wrong. This isn't merely about misplaced technological optimism. It's about prying into the private lives of young girls through racist, classist datasets, powered by state–corporate collusion, all while disregarding fundamental principles of privacy, ethics, and security. Salta's so-called Technological Platform of Social Intervention offers a chilling parable of how surveillance and oppressive algorithms can be weaponized against the vulnerable—here a group of Brown and Black girls.

1

While this is not the end of the story, the focus of this book is one around solutioning, so I will shift focus, as promised, away from the outrage and toward what lies within our control to help mitigate the oppressive AI-powered design systems that are weaponized against women and girls. Accordingly, the goal of this book is to explore how we counter such nefarious practices that infringe on the rights of women and girls.

As I conducted my own doctoral research on how AI-powered data design systems have been weaponized (Lubin & Harris, 2024), this Argentinian example of sexist data practices was just one of the many moments that jolted me awake to a new reality—a world of female datafication that eerily mirrors *The Handmaid's Tale* (Atwood, 1985) in its dystopian representation of a subjugated female class but now in digital form. But this practice is not new at all when we consider the deep histories of women and girls in the transatlantic slave trade, where Black women's bodies were not only commodified for labor but also subjected to meticulous recordkeeping and surveillance that reinforced racial and gendered hierarchies (Hartman, 1997; Morgan, 2004). Enslaved Black women were counted, categorized, and systematically controlled through ledgers, insurance records, and plantation logs—an early form of biometric data collection that foreshadowed today's AI-driven systems of classification and exploitation (Berry, 2017; Browne, 2015). Now, as we move from analogue to digital, feminists and activists like Paz Peña and Joana Varon (2021) have also begun answering the hard questions these digital systems raise, crafting a critical feminist framework that challenges the status quo of AI ethics.

While many ethicists focus on transparency, privacy, and safety, feminists ask more fundamental questions:

Whose power does it serve?

Who profits?

Who loses?

This book explores those deeper questions and, more importantly, how we take action. It moves beyond merely identifying the problems inherent in these AI-powered systems to envisioning and implementing transformative solutions. Granted, the speed of AI-powered development is unquestionably fast, so we must be equally swift and intentional in our response. This book challenges us not just to critique but to build—to reimagine data systems that center equity, justice, and humanity. The question is no longer whether AI will shape our future but, rather, how we will shape AI to serve us all.

Grounded in feminist principles, the book offers a road map for dismantling oppressive power structures embedded in technology and building equitable, inclusive systems that center the voices of those historically marginalized. Through case studies, frameworks, and actionable strategies, it challenges readers to imagine a future where AI serves collective liberation rather than corporate or state control.

I would have liked to begin by stating that women's rights and civil liberties are under attack. However, one need only look to the Argentinian Salta province case, alongside the ongoing assaults on women's rights, to see the undeniable truth. The reality is that these attacks are neither new nor isolated—they are part of a long, insidious history of controlling women's bodies through policy, technology, and systemic oppression. From the forced reproductive labor of enslaved Black women to contemporary digital surveillance and data-driven discrimination, the mechanisms may have evolved, but the underlying logic of control remains the same.

The recent reversal of *Roe v. Wade* (1973; see *Dobbs v. Jackson Women's Health Organization*, 2022) and the resurgence of a political party that championed this rollback further underscore the gravity of the situation. We are in uncharted territory where decades of hard-won rights are being systematically dismantled under the guise of legal and technological progress. Yet, while these challenges demand recognition, this book is not merely an autopsy of our losses—it is a blueprint for action. Rather than lingering on despair, it shifts the focus toward resistance, innovation, and the transformative solutions needed to reclaim autonomy in an era where digital systems, legislation, and policy are being weaponized against marginalized communities.

These anecdotes are just a few examples of the dangerous intersection between technology, patriarchal control, and surveillance—a convergence that demands we look deeper. And we cannot forget that these AI-powered systems are not just technical artifacts; they embody the biases and power dynamics of the societies that create them. As D'Ignazio and Klein (2020) argue in their groundbreaking book *Data Feminism*, data systems are never neutral. They emphasize how the framing, collection, and application of data often reinforce existing inequalities, particularly along lines of gender, race, and class.

It's time we ask the hard questions before the answers are dictated for us:

> *What values do these systems encode?*
>
> *Whose perspectives do they prioritize, and whose do they exclude?*

Feminist scholarship reveals that without critical interrogation, the same technologies heralded as tools of progress can become mechanisms of oppression (D'Ignazio & Klein, 2020). As we stand at the crossroads of rapid technological expansion, we must recognize that digitization is not a neutral force—it reflects the biases, priorities, and power structures of those who design and control it. If left unchecked, these systems will not only replicate historical injustices but also amplify them at an unprecedented scale. The question is not just whether we can trust these technologies but whether they can be reimagined to serve justice rather than perpetuate harm. The future of data is being written now, and we must decide if we will be passive subjects in its design or if we will take agency in shaping an equitable digital world.

GENDER INEQUITIES IN A DIGITIZED WORLD

As digital tools become more pervasive, their design and implementation often reflect the biases and inequities of the societies that create them. Women, in particular, bear the brunt of these systemic flaws, as they are disproportionately affected by discriminatory algorithms, data exploitation, and limited access to digital resources. To understand the broader implications, we must explore how these disparities manifest across economic, social, and technological domains. These digital inequities do not exist in isolation—they intersect deeply entrenched economic and social disparities, further entrenching barriers to equality. The same systemic biases that shape discriminatory algorithms also underpin wage gaps, labor exploitation, and economic disempowerment. To fully grasp the scope of these challenges, we must examine how digital marginalization compounds existing economic inequalities, reinforcing a cycle of exclusion that limits women's opportunities and financial independence.

Women constitute approximately half of the global population yet continue to face significant economic disparities, notably the persistent gender pay gap. Globally, women earn about 84 cents for every dollar earned by men, indicating a 16% wage gap (Fleck, 2024). In the United States, this gap is slightly wider; in 2023, women earned 83% of what men earned, a decline from 84% in 2022 (AP News, 2023). The gender pay gap varies across countries; for instance, in South Korea, men out-earned women by 29.3% in 2023, while in Japan, the gap was 22% (Fleck, 2024). Despite some progress, the global gender gap remains significant. The World Economic Forum's *Global Gender Gap Report 2024* indicates that the global gender gap is 68.4% closed, reflecting only a 0.3 percentage point improvement from the previous year (Pal et al., 2024). These statistics underscore the ongoing challenges women face worldwide in achieving economic equality. Data are not neutral—they reflect the biases of those who create them, and as depicted in Figure I.1, we are reminded of the deep inequities that frame our ontological existences as women.

This wage disparity across gender not only affects individual earnings but also has broader economic implications, including reduced household income and diminished economic growth. More critically, it undermines women's autonomy and civil liberties by restricting their financial independence, limiting access to essential resources such as health care and education, and reinforcing systemic power imbalances. Economic inequality is not just about pay—it is about control, agency, and the ability to fully participate in society without financial constraints dictating women's choices.

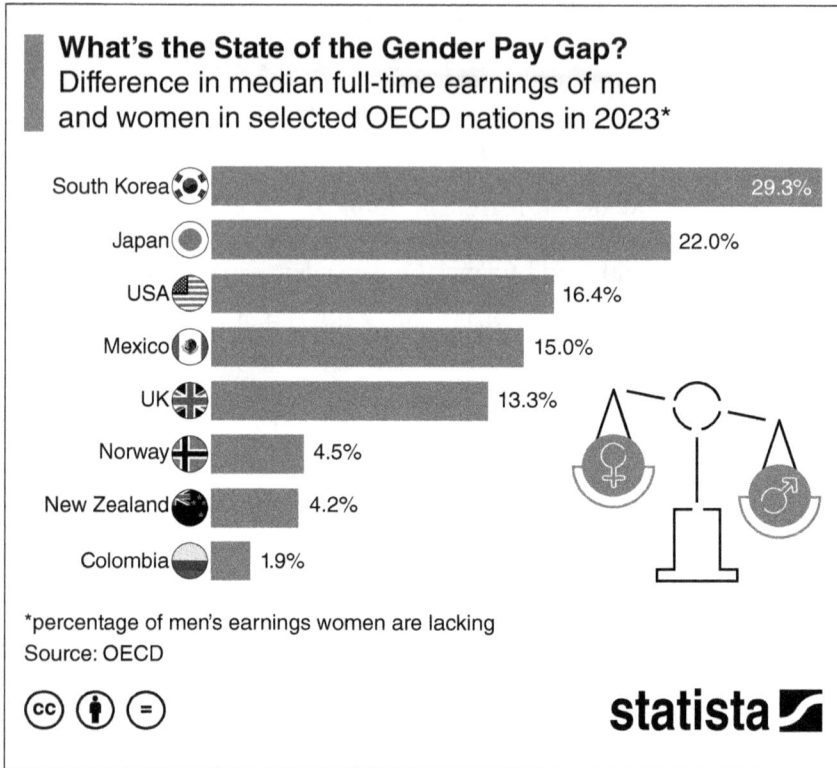

What's the State of the Gender Pay Gap?
Difference in median full-time earnings of men
and women in selected OECD nations in 2023*

Country	Gender Pay Gap
South Korea	29.3%
Japan	22.0%
USA	16.4%
Mexico	15.0%
UK	13.3%
Norway	4.5%
New Zealand	4.2%
Colombia	1.9%

*percentage of men's earnings women are lacking
Source: OECD

statista

Figure I.1　Gender Pay Gap Across OECD Countries in 2023

Source: *What's the State of the Gender Pay Gap?*, by A. Fleck, November 13, 2024, Statista (https://www.statista.com/chart/13182/where-the-gender-pay-gap-is-widest/). CC BY-ND 3.0.

AUTONOMY, PRIVACY, AND LIBERTIES

In the realm of data privacy and civil liberties, the situation is equally concerning. As technology becomes increasingly integral to our daily life, the collection and utilization of personal data have profound implications for privacy and autonomy.

This intersection of technological advancement and entrenched societal inequities raises critical questions about the ethics and fairness of AI systems. While data privacy and civil liberties are central to discussions about individual autonomy, these concerns often intersect with deeper, systemic biases that have historically marginalized certain groups, particularly women. As algorithms increasingly shape decision-making processes in health care, employment, and social policy, understanding how historical inequalities manifest in contemporary AI systems becomes essential for ensuring equitable outcomes. This leads to a broader examination of how long-standing gender disparities are perpetuated and amplified in the digital age.

It is imperative to recognize that the historical context surrounding the treatment and representation of women's bodies has significantly shaped the biases embedded in contemporary AI systems.

Barocas and colleagues (2020) provide a meticulous analysis of how traditional gender stereotypes are systematically encoded into machine learning algorithms through biased training data. Their work demonstrates that such data, reflective of entrenched societal norms, not only perpetuate historical inequalities but also legitimize them under the guise of technological neutrality. They emphasize the ethical and technical challenges in mitigating these biases, calling for more equitable approaches to data representation and algorithmic design.

Perez (2019), in her groundbreaking exploration of systemic data bias, underscores the pervasive exclusion of women in data collection processes across industries. She illustrates how this exclusion leads to the development of technologies that are not only less effective but often actively harmful to women, particularly in areas such as health care, urban planning, and workplace safety. Her work serves as a stark reminder that data are not inherently objective; rather, they reflect the priorities and blind spots of those who collect the data, making the case for gender-inclusive practices in data governance and technological innovation.

Safiya Noble's *Algorithms of Oppression* (2018), for example, underscores how search engines' algorithms perpetuate historical stereotypes about women, attributing this issue to the lack of diverse representation in AI development. Her work opens up critical discussions about the female and race effect in AI, shedding light on how biased data and homogenous development teams lead to technologies that not only fail to represent women equitably but also actively reinforce harmful stereotypes.

Relatedly, I had the honor of cohosting Dr. Noble (2021) during my time as a graduate student at the University of Waterloo, Canada, as part of the International Symposium on Technology and Society in a virtual fireside conversation. Dr. Noble shared the deep, personal weight of her research into oppressive AI systems and the relentless burden of uncovering how these technologies reinforce racial and gender injustices. As a Black woman, she spoke of the emotional toll of confronting a system that not only marginalizes people who look like her but also demands that she continually prove its harm, even as she experiences it firsthand. This moment remained with me as I was myself about to complete this journey and was discovering much the same and grateful, in that moment, that I was not alone in this pursuit of knowing more about the many systematic biases that plague us, not only as Black people but even more so as women. Her keynote address also provided a critical lens on the intersection of diversity, equity, and inclusion in AI and technology design, generally. In that keynote (Noble, 2021), she situated her work within a historical framework, acknowledging the contributions of past activists and researchers who have paved the way for interrogations of bias and inequality in technological systems. She also simultaneously underscored the accelerating proliferation of negligent technologies over the past two decades, emphasizing their profound influence on societal structures defined by gender, race, and sex.

In this poignant exchange during the keynote (Noble, 2021), one of my moderation prompts elicited Dr. Noble's articulation of two major concerns: the lack of regulatory oversight in technology and the hyper-reductionist approach of the industry, which attributes bias solely to computer code. She went on to dismantle this reductionist narrative, asserting that centuries of systematic discrimination have been encoded into technological systems, rendering these biases opaquer and more entrenched, sometimes invisible to those who do not fully understand the black box mystique of AI. She also framed this as a violation of civil and human rights, underscoring the urgent need to address these inequities at structural and systemic levels.

Despite the challenges inherent in shifting from paradigms of surveillance and extraction to one of justice, Dr. Noble (2021) concluded the keynote with a message of hope. She reminded the audience, comprising mostly graduate students (including a subset of engineering grad students) and faculty, of the collective progress made and the shared responsibility for addressing these issues.

Her vision encourages reimagining engineering practices, inclusive of design, through the lenses of history, empathy, and care, offering a pathway toward creating more equitable technological futures. This conversation not only deepens our understanding of the intersection of race, AI, and technology but also emphasizes the necessity of interdisciplinary approaches and collaborative efforts to foster accountability and justice in technological systems. While this conversation highlighted actionable steps toward justice and accountability, it also recalled the deep histories of Black bodies in White spaces since the days of the enslavement of African bodies (Noble, 2021).

And as we have come to realize, when we factor race into the technology dialogue, the historical exploitation of Black women's bodies provides a critical lens for understanding the biases inherent in modern AI systems, a topic worth discussing briefly for foundational context of why these practices ensure so deeply in American histories. Frances M. Howell (2023) explores the harrowing medical abuse endured by Black women during slavery, illustrating how their bodies were exploited as tools for medical experimentation and advancement, often without consent or regard for their humanity. This exploitation not only dehumanized these women but also laid the foundation for systemic disparities in health care that continue to affect marginalized communities today. The legacy of these abuses is evident in the mistrust many Black women feel toward medical institutions and the persistent disparities in maternal mortality, access to care, and pain management.

The historical exploitation and abuse of Black individuals within medical contexts have fostered a deep-seated mistrust among Black women toward health care institutions. This mistrust is compounded by persistent disparities in maternal mortality, access to care, and pain management. According to a report from the U.S. Centers for Disease Control and Prevention (Hoyert, 2023), Black women in the United States are three to four times more likely to die from pregnancy-related causes than White women, a disparity that transcends socioeconomic status and education levels.

Additionally, studies have documented that Black patients are less likely to receive appropriate pain management compared to their White counterparts, reflecting systemic biases in medical treatment (Strand et al., 2021). This knowledge came to surface, especially during the COVID-19 pandemic, along with the all-too-common tales of the medical inequities faced by Black and Brown people. Recalling the horrific tale of Dr. Susan Moore, a Black physician who documented her own mistreatment in the health care system before dying of COVID-19, underscores the devastating consequences of medical bias. Despite being a doctor herself, Dr. Moore was dismissed, denied proper pain management, and accused of drug-seeking behavior—experiences that mirror the systemic racism countless (Wixson, 2021). Black patients endure. Her tragic death became a stark reminder that medical credentials do not shield Black individuals from discrimination and that the deep-seated mistrust in health care institutions is rooted in lived realities, not paranoia. The COVID-19 pandemic merely exposed what has long been true: Racial disparities in health care are not incidental; they are systemic, persistent, and deadly. It is a tale also captured in the germinal work of Nikole Hannah-Jones and The New York Times Magazine's *1619 Project* (2021). These ongoing inequities underscore the critical need for systemic reforms to build trust and ensure equitable health care outcomes for Black women, whose data, too, are implicated in the design of these systems.

By exposing these historical injustices, Howell (2003) emphasizes the pressing need to confront the structural inequities in health care that sustain cycles of marginalization and erasure. These narratives are not merely historical artifacts; they serve as a rallying cry to reimagine and reconstruct a health care system founded on principles of equity, **informed consent,** and universal dignity.

With the increasing integration of technology into health care delivery, the urgency to address and dismantle codified analog biases embedded within these systems has never been greater. This intersection of technology and care demands deliberate action to ensure that innovation does not perpetuate the injustices of the past but instead fosters a more inclusive and just future.

In *Race After Technology: Abolitionist Tools for the New Jim Code*, Ruha Benjamin (2019) deepens the discourse on systemic biases in technology, extending her analysis to AI-driven surveillance systems. Benjamin illustrates how these technologies disproportionately target Black women, reinforcing a historical legacy of control and dehumanization rooted in colonialism and slavery. She argues that such systems are not merely flawed or neutral tools but are active participants in the reproduction of structural inequities, subtly legitimizing the surveillance and policing of marginalized communities under the guise of technological progress. Benjamin's work calls for an abolitionist reimagining of these systems to dismantle oppressive frameworks rather than reinforce them.

Similarly, Joy Buolamwini and Timnit Gebru, in their essay "Gender Shades" (2018), provide a pivotal empirical analysis of racial and gender biases embedded in facial recognition technologies. Their research reveals significantly higher error rates for Black women, highlighting the ways in which technological systems reflect and perpetuate historical practices of exclusion and marginalization. They demonstrate how these biases are not accidental but are a direct consequence of imbalanced training datasets and the prioritization of normative identities in algorithmic design. By exposing these disparities, Buolamwini and Gebru emphasize the urgent need for equitable practices in the development of AI, advocating for accountability and inclusivity to ensure these tools do not replicate the injustices of the past.

Similarly, Charlton McIlwain's *Black Software: The Internet and Racial Justice, From the AfroNet to Black Lives Matter* (2019) reveals how digital infrastructures have historically marginalized Black voices while perpetuating discriminatory practices. Together, these works illustrate how the misuse of data disproportionately harms Black women, reinforcing systemic inequalities across societal domains (Benjamin, 2019; Buolamwini & Gebru, 2018; McIlwain, 2019).

The surveillance and representation of Black female bodies have long been deeply entrenched in colonial and apartheid legacies, with African female scholars critically analyzing these enduring issues. Desiree Lewis (2001) critiques dominant narratives that frame African women through sexualized and dehumanizing lenses, emphasizing the need to reclaim agency and construct narratives that resist oppressive stereotypes. Lewis highlights the ways in which African women have been historically objectified, calling attention to the importance of reshaping these narratives to prioritize dignity and autonomy.

Pumla Dineo Gqola (2015) builds on these ideas by examining the pervasive violence and hypervisibility of Black women, demonstrating how historical constructs of racial and gendered oppression continue to shape the policing of their bodies. Gqola argues that the deeply entrenched histories of systemic violence against Black women have contemporary manifestations, making it imperative to critique and resist these practices. She brings into focus how the colonial and apartheid histories of surveillance perpetuate cycles of oppression.

Gabeba Baderoon (2014) extends this discourse by exploring the specific impact of racialized and gendered surveillance practices on Muslim women in South Africa, connecting these practices to broader mechanisms of control and exclusion. Baderoon's analysis links the historical representation of women of color to the surveillance tactics used to marginalize them further, providing a lens to understand how colonial histories are carried into present systems of governance and societal norms.

By integrating these historical perspectives into the design and governance of data systems, we can foster more just and inclusive representations that prioritize the dignity and agency of Black women (Baderoon, 2014; Gqola, 2015; Lewis, 2001).

The Enduring Legacy of Sarah Baartman: Colonial Exploitation and the Commodification of Black Women's Bodies

The case of Sarah Baartman, often referred to as the "Hottentot Venus," is a harrowing example of the historical exploitation and dehumanization of Black women's bodies. Born in South Africa's Khoikhoi community, Baartman was coerced into traveling to Europe in 1810 under the pretense of employment opportunities. Once there, she was paraded as a human spectacle in Britain and France, with her physical features—particularly her large buttocks—objectified and fetishized by colonial audiences who regarded her as an exotic curiosity (as illustrated in Figure I.2). Her treatment reflected the deeply ingrained colonial ideologies that positioned Black women as "other," reducing them to mere bodies for entertainment and scientific scrutiny.

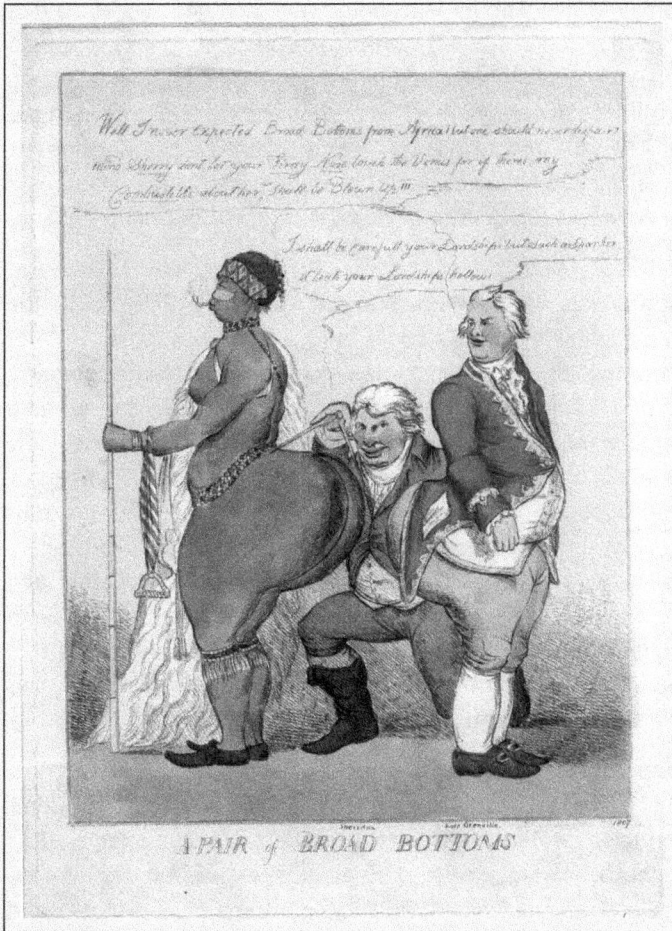

Figure I.2 Caricature of Sarah Baartman by William Heath (1810)

Source: "*A Pair of Broad Bottoms*," *a Caricature by William Heath From 1810*, uploaded by user Dencey on August 10, 2021, to Wikimedia Commons (https://commons.wikimedia.org/wiki/File:A_Pair_of_Broad _Bottoms.jpg). In the public domain.

(Continued)

(Continued)

Baartman's exploitation did not end with her death in 1815. Her remains, including her skeleton, brain, and genitalia, were dissected and preserved by French scientist Georges Cuvier, who used her body as evidence to support racist theories of human difference and inferiority. Her body parts were displayed at the Musée de l'Homme in Paris for over a century, perpetuating her objectification even in death. It was not until 2002, after years of advocacy, that her remains were returned to South Africa for a proper burial.

Baartman's life and posthumous treatment underscore the intersection of race, gender, and systemic oppression. Her story reveals the historical commodification of Black women's bodies within colonial and patriarchal structures, serving as a chilling reminder of the ways in which science and society have collaborated to dehumanize and marginalize. This legacy continues to inform contemporary systems of oppression, particularly in the realms of representation and datafication, where Black women's bodies are still subjected to reductive and harmful narratives. The case of Sarah Baartman compels us to critically examine and challenge the structural inequities that persist in modern institutions, ensuring that dignity and humanity are prioritized in all contexts.

Her story is not just a relic of the past but a reflection of how Black women's bodies have historically been reduced to commodities or curiosities, devoid of agency or humanity. This legacy of exploitation and objectification has echoes in modern AI systems, where data and algorithms often replicate and reinforce historical biases. By failing to represent the full spectrum of humanity equitably, these systems risk perpetuating the same dehumanization that women like Baartman endured. Her story serves as a vital historical touchstone for interrogating the ways race and gender bias continue to shape emerging technologies.

Notably, the objectification and commodification of women—particularly the reduction of their womanhood to biological processes such as menstrual cycle tracking—has been encased within technological systems. These systems position such intimate aspects of women's lives as phenomena to be monitored, modeled, and monetized, perpetuating historical patterns of control and exploitation.

Building on these historical insights, Dr. Nour Naim (2023) analyzes the evolving dynamics of AI-driven surveillance technologies in her essay, "Women in the Era of Artificial Intelligence: Increased Targeting and Growing Challenges." She reveals how such systems exacerbate existing inequalities, disproportionately targeting women and amplifying long-standing trends of control, marginalization, and oppression. Naim highlights the ways these technologies embed systemic biases into their design and deployment, reinforcing patriarchal structures and deepening the vulnerabilities of women, particularly those from marginalized communities.

She further critiques how these technologies, often implemented without accountability or consideration of their social impact, amplify power imbalances by subjecting women to heightened scrutiny and eroding their privacy (Naim, 2023). Naim's (2023) work underscores the broader consequences of algorithmic surveillance, showing how these systems replicate and intensify patriarchal structures, undermining women's autonomy and freedom. This analysis calls for a reevaluation of AI governance, emphasizing the need to address the structural inequities underpinning these technological systems to prevent perpetuating cycles of control and marginalization.

The discourse surrounding AI in health care has increasingly highlighted systemic biases that disproportionately affect women's health. As noted by Lau (2024), the lack of gender parity in medical data reflects entrenched discrimination, both historically and in contemporary contexts, resulting in significant algorithmic gender bias. This gender data gap not only permeates AI technologies but extends into health care protocols, adversely influencing women's treatment outcomes and health rights. Lau underscores the urgency of addressing these disparities, emphasizing the need for comprehensive analysis of legal frameworks that govern AI-driven health care across jurisdictions such as the European Union, the Council of Europe, and the United Kingdom. The investigation reveals that while supranational AI regulations exist, they often fall short of explicitly safeguarding women's fundamental health rights. Accordingly, Lau advocates for integrating data feminism, rooted in intersectionality theory, as a mechanism to foster gender equity and counteract **algorithmic bias** in health care systems.

D'Ignazio and Klein (2020), in their feminist critique, emphasize the systemic exclusion of women's perspectives in data collection and AI development, proposing actionable frameworks to create more equitable systems. Their approach challenges the foundational structures of data science and AI, advocating for intersectional methods that center marginalized voices and actively dismantle ingrained biases. By reimagining how data are collected, interpreted, and applied, they offer a blueprint for developing systems that reflect diverse experiences and promote social justice.

Similarly, Criado Perez (2019) explores how systemic gender bias in data collection perpetuates inequality across health care, technology, transportation, and public policy. Criado Perez demonstrates how the exclusion of women from datasets lead to products, systems, and environments that neglect half the population—often with dangerous or fatal consequences. Her work, like that of others, re-echoes the findings of Buolamwini and Gebru, whose study "Gender Shades" (2018) exposed the failure of AI systems to accurately recognize Black individuals and darker-skinned people due to unrepresentative datasets. This alignment has led some critics to contend that Criado Perez's contributions build on well-established feminist and technological critiques, raising questions about the novelty of her arguments. Additionally, while Criado Perez acknowledges the importance of intersectionality, some critiques suggest that the book's primary focus on gender may inadvertently overlook the compounded biases experienced by marginalized women, including women of color, women with disabilities, and LGBTQ+ communities.

This evolving dialogue signals a broader imperative for gender-sensitive legal and technological interventions to ensure equitable health care outcomes, reinforcing the necessity for interdisciplinary collaboration between AI developers and designers, health care professionals, and policymakers. Rather than reinforcing this critique, it is crucial to recognize the complex nature of bias, particularly in AI, and the necessity of addressing these issues within distinct contexts. Health care, for instance, emerges as a critical area where biased AI design poses significant risks to women's health, disproportionately affecting marginalized communities and prompting a broader reevaluation of womanhood in data-driven systems. While some critique *Invisible Women* (Criado Perez, 2019) for its narrow scope, I argue that centering women as a collective—while explicitly addressing the nuances of intersectionality—strengthens the broader pursuit of justice. Women's rights, as a facet of human rights, must encompass those who experience intersecting forms of discrimination.

As AI and known, as well as unknown, emerging technologies continue to reflect and amplify societal biases, the urgency of this conversation intensifies. Although awareness of AI's discriminatory patterns is increasing, there is a pressing need for tangible interventions, accountability mechanisms, and proactive strategies to dismantle the structures enabling these inequities. In this context, **ethotic heuristics**—a framework that centers ethics, social context, and lived experience in technological design—represent a pivotal step toward addressing the hidden inequities embedded in AI-powered data design. This framework not only surfaces systemic biases but also equips practitioners with practical tools to directly confront and mitigate these issues. By embedding accountability and auditability into the AI design process, ethotic heuristics facilitate the development of more ethical and inclusive systems, ensuring that bias is actively corrected rather than passively acknowledged.

Collectively, these works—foregrounding the intersections of race, gender, and history—underscore the imperative to confront gender disparities in AI to foster more equitable technologies that serve all communities. While these inequities disproportionately affect Black and Brown populations, the insights and solutions presented are essential for cultivating broader equitable design practices that uplift all marginalized groups and contribute to the creation of fair and just technological systems.

THE INTERSECTION OF TECHNOLOGY, REPRODUCTIVE RIGHTS, AND HEALTH CARE IN A POST-*ROE* ERA

The reversal of *Roe v. Wade* (1973) in the June 2022 *Dobbs* case has triggered profound implications for reproductive health care, data privacy, and medical ethics across the United States. This legal shift not only restricts access to abortion care but also exposes women—particularly from marginalized communities—to heightened surveillance, prosecution, and systemic inequities.

Health Care Consequences and Maternal Mortality

Since the *Roe* reversal, maternal deaths have risen sharply, cor-relating with restricted abortion access in certain states. A study by Declerque et al. (2022) found that maternal mortality rates in states with strict abortion laws were 62% higher than in states with broader reproductive health care access. This alarming trend exemplifies how legal and political restrictions translate into tangible health risks.

Tragic cases such as those of Porsha Ngumezi and Amber Nicole Thurman illustrate these dire consequences. Ngumezi, a 35-year-old from Texas, died from hemorrhaging during a mis-carriage when doctors, fearing legal repercussions, hesitated to perform a dilation and curettage (Presser & Surana, 2024). Similarly, Thurman died from septic shock in Georgia after delays in her incomplete medication abortion, a case that has raised concerns about the impact of restrictive laws on emer-gency medical care (Branstetter, 2024; Surana, 2024). These stories stand as stark reminders of the human cost behind restrictive policies. By documenting their names, we honor their memory and spotlight the disproportionate impact on Black women, while acknowledging that these risks extend across all demographics.

Erosion of Medical Ethics and Patient Trust

The chilling effect of restrictive abortion laws reaches beyond mortality statistics, influencing the decisions of health care pro-viders. Physicians, bound by the Hippocratic Oath to prioritize patient welfare, increasingly find themselves caught between ethical obligations and legal constraints (Congressional Record, 2017). Fear of prosecution discourages timely medical intervention, directly compromising patient care and undermining trust in the health care system.

The erosion of ethical principles not only places patients at risk but also signals a departure from foundational medical values. This departure disproportionately affects women from marginalized communities, further exacerbating health disparities and reinforcing systemic inequities.

Technology as a Double-Edged Sword

In parallel, technology—often seen as a vehicle for empowerment—has become a tool of surveillance in the post-*Roe* landscape. AI-powered health applications, such as period trackers and fertility monitors, are marketed as empowering tools but risk exposing users to privacy violations. As Spector-Bagdady and Mello (2022) highlight, law enforcement agencies have accessed electronic health records and period-tracking app data to prosecute individuals seeking abortions.

A notable case in Nebraska exemplifies this troubling trend, where prosecutors obtained private Facebook messages between a mother and her 17-year-old daughter, resulting in abortion-related charges ("Why Data Privacy Is a Concern," 2022). This case marked the first known instance of digital communication being used in such legal proceedings.

Similarly, Cox (2022) and Lapperruque et al. (2022) warn of the increasing vulnerability of digital footprints, from search histories to geolocation data, emphasizing the urgent need for robust data protections. Without clear regulatory frameworks, private reproductive data remain exposed to exploitation.

Intersection of Law, Technology, and Reproductive Justice

Legal experts stress the importance of revisiting data governance frameworks to safeguard reproductive health information. The reversal of *Roe* underscores the necessity of ethical AI design and regulatory measures that prioritize privacy, ensuring that personal health data cannot be weaponized in legal or political contexts.

As highlighted by Lau (2024), AI-driven health care systems already exhibit gender biases rooted in historical data gaps. The intersection of these biases with reproductive rights presents new ethical dilemmas, necessitating comprehensive policy responses across jurisdictions, including the European Union, Council of Europe, and United Kingdom. Lau advocates for integrating data feminism—a framework championed by D'Ignazio and Klein (2020)—to challenge existing power structures and mitigate algorithmic bias in health care.

Data Feminism and the Path Forward

In *Data Feminism*, D'Ignazio and Klein (2020) argue that data are never neutral; they reflect societal inequities unless deliberately countered. Their framework emphasizes the need for inclusive data governance, **algorithmic accountability**, and equitable representation in AI development. In the post-*Roe* era, where women's reproductive health data face unprecedented scrutiny, these principles take on new urgency. Implementing data feminism requires actionable steps, all accounted for with ethotic heuristics:

- **Consent-Driven Data Collection:** Ensuring individuals have control over how their reproductive health data are gathered and shared

- **Algorithmic Accountability:** Auditing AI systems to identify and rectify gender biases

- **Privacy-Centered Governance:** Enforcing stricter regulations around health-related data, protecting them from unauthorized access or surveillance

By embedding ethical considerations into technological frameworks, we can resist the surveillance state and safeguard reproductive rights. Without such measures, the digital landscape risks perpetuating harm, further disenfranchising women at a time when their rights are under assault (Eubanks, 2018).

The intersection of technology, reproductive rights, and health care also serves to highlight the urgent need for systemic change. As *Roe*'s reversal reshapes the legal and ethical terrain, protecting women's health data and addressing systemic inequities must remain at the forefront of policy and technological development. By embracing principles of data feminism and holding institutions accountable, we can forge a more equitable future where reproductive health care is safeguarded, not surveilled.

THE CHANGING FACE OF HEALTH CARE

In *The Doctor and the Algorithm: Promise, Peril, and the Future of Health AI*, S. Scott Graham (2022) examines the integration of AI into health care, focusing on the collaborative dynamics between human clinicians and AI systems. Graham explores how AI can enhance diagnostic accuracy and treatment planning, while also addressing ethical concerns such as data privacy, algorithmic bias, and the potential for AI to inadvertently reinforce existing health care disparities. He emphasizes the importance of maintaining human oversight in AI-assisted medical decisions to ensure patient-centered care. Graham further advocates for a balanced approach that leverages AI's capabilities without compromising the essential human elements of empathy and ethical judgment in medicine. However, his vision for human oversight in health care is overshadowed by the reality of commoditized AI systems, which often exploit the female body as a locus for profit within Big Tech's frameworks. This dynamic further marginalizes women, stripping agency from an already vulnerable demographic and highlighting the need for a more equitable integration of AI in health care.

Women are vulnerable to data practices that may infringe upon their civil liberties, leading to issues such as targeted discrimination and exploitation. Ensuring that data governance frameworks respect and protect the rights of women is essential to uphold their civil liberties and promote equitable participation in the digital economy.

When it comes to women's health care data, the stakes are even higher. From reproductive health and fertility apps to medical research databases, the ways in which women's health care data are collected, stored, and analyzed have far-reaching implications for their privacy, health outcomes, and autonomy. The absence of ethical oversight in these practices risks weaponizing data against women, leading to issues like predictive policing in reproductive health care or biased medical algorithms that fail to consider diverse biological markers.

Scholars like Benjamin (2019) emphasize the dire consequences of embedding societal inequities into technological systems without rigorous critical examination, underscoring the urgent need for heuristic frameworks to guide their design and implementation. Her work, along with that of other prominent scholars, highlights the critical importance of adopting **intersectional data practices**—attentive to race, gender, and socioeconomic factors—to mitigate the disproportionate harm these systems often inflict on marginalized communities. Benjamin's analysis sheds light on the underrepresentation and misrepresentation of Black individuals in digital spaces, urging the development of inclusive technological systems that actively dismantle, rather than perpetuate, systemic injustices.

To meet this challenge, ethotic heuristics must become central to AI governance. Without deliberate, justice-oriented methodologies, we risk not only codifying historical biases but accelerating them at an unprecedented scale. These heuristics offer a necessary counterbalance, ensuring that technological advancements are evaluated not just for efficiency but for their ethical and social impact. As the reach of AI expands, so too must our commitment to developing frameworks that foreground human dignity, equity, and accountability in the digital age, highlighting that dangers that lurk at the intersection of sex and technology.

SEX AFTER TECHNOLOGY: ETHOTIC HEURISTICS AS A COUNTERMEASURE TO THE MISREPRESENTATION OF PEOPLE

Benjamin's *Race After Technology* (2019) served as a foundational influence on the essay that ultimately sparked the development of ethotic heuristics. In her work, Benjamin introduces the concept of the New Jim Code, highlighting how racial biases are not just embedded in digital systems but actively reinforced through automated processes that masquerade as neutral or objective. Her analysis of the racialized dimensions of algorithmic governance deeply informed my coauthored essay with my academic supervisor, computational rhetorician Dr. Randy Allen Harris, titled "Sex After Technology: The Rhetoric of Health Monitoring Apps and the Reversal of *Roe v. Wade*," published in *Rhetoric Society Quarterly* (Lubin & Harris, 2024). In this work, we critically examine the unsettling intersection of AI technologies with the rise of Christo-fascist movements in the United States, exposing how digital surveillance, data-driven discrimination, and algorithmic governance are being weaponized to reinforce patriarchal and authoritarian ideologies—echoing the ways the New Jim Code perpetuates racial inequities under the guise of technological progress.

This intersection of sex and technology poses profound threats to women's health, particularly in the wake of the reversal of *Roe v. Wade*. These risks are amplified by pervasive and well-documented privacy violations perpetrated by major technology corporations and organizations, often with the tacit or explicit sanction of state actors, including law enforcement. These entities exploit algorithmically mined data to construct "user models," which are then used to surveil, control, or penalize women, further entrenching systemic inequities and eroding their personal autonomy.

These models—aggregations of fragmented data points—are frequently sold or shared without genuine informed user consent, raising urgent ethical and legal concerns in an era of increasing technological intrusion into personal health and reproductive autonomy. Such algorithmic constructions, or what I term **algorithmic ethopoeia**—the mathematized representations of individuals charted as commodified matrices for commercial gain, political maneuvering, or law enforcement purposes—do more than exploit user data. They reflect a fundamental betrayal of privacy and autonomy, while also contributing to a profound erosion of individual agency.

Building on abolitionist frameworks for resisting intersectional racism, as articulated within critical race technology studies, and incorporating principles from data feminism, we (Lubin & Harris, 2024) propose six categories of design heuristics aimed at safeguarding and honoring the ethopoeic integrity of women. These combined approaches confront the structural biases embedded in data systems, calling for the disruption of automated characterizations that perpetuate and amplify societal inequities. By reframing data from an immutable categorization of individuals to a dynamic construct shaped by power dynamics and diverse perspectives, we advocate for ethical and inclusive design practices that actively dismantle mechanisms of dehumanization and foster equity.

The foundations of critical race studies, technology studies, and gender studies offer a vital precursor to discussing what I call ethotic heuristics in technology.

I define ethotic heuristics as a framework at the intersection of rhetoric and design advocating for ethical principles and practical guidelines aimed at counteracting the misrepresentation and dehumanization of individuals in technological systems. Rooted in intersectional and feminist perspectives, these heuristics emphasize the moral responsibility of developers and policymakers to design systems that recognize and respect the dignity, agency, and diversity of all users. By addressing the power imbalances and biases inherent in data collection, modeling, and application, ethotic heuristics aim to promote equity, inclusivity, and justice in technology.

Ethotic heuristics, therefore, function as a meta framework that advocates for the elimination of mechanisms enabling machines to characterize humans. By prioritizing dignity, agency, and ethical accountability in technological design, our heuristic framework addresses the complex political and legal entanglements that increasingly shape and constrain the development and application of such systems. These heuristics, we believe, are particularly critical for ensuring ethical technological design in women's health amid the challenges of a post-*Roe* landscape (Lubin & Harris, 2024).

Here, in this book, I expand on these ethotic heuristics, with specificity for practical use, in the hopes of further informing a more robust auditing framework, which has become critical in today's sociopolitical and technological landscape and what that will mean for the civil liberties of women. By prioritizing transparency, accountability, and inclusivity in data practices, we can work toward creating systems that empower women rather than marginalize them.

Algorithmic Ethopoeia and the Need
for Ethotic Heuristics in Health Care

As the completion of this book drew near, it became essential to revisit this section to incorporate a significant and timely news development that has captured widespread public attention: the tragic and homicidal death of UnitedHealthcare CEO Brian Thompson. This event, coupled with the ongoing state of health care in the country, demanded further exploration. Additionally, a closer examination of the company's claims approval process revealed a critical detail: the AI algorithm known as nH Predict (Mello & Rose, 2024). Designed to make coverage decisions, this AI system exemplifies the concept of algorithmic ethopoeia—where AI-driven systems, such as the company's claims processor, create characterizations of individuals that directly influence high-stakes decisions, such as the approval or denial of insurance claims.

In this case, the algorithm was alleged to have a 90% error rate, leading to wrongful denials of medically necessary rehabilitation services for Medicare patients (Mole, 2023). By cross-referencing patient records with aggregated data from other patients, the algorithm prioritized pattern recognition over individualized care, exposing the ethical and practical dangers of such characterizations (Mello & Rose, 2024). This inquiry is especially urgent in an era where generative AI is reshaping industries, profoundly influencing gender dynamics in myriad spaces like employment and health care.

At the heart of this discussion lies the critical need to interrogate how AI systems either perpetuate or challenge existing inequalities. Central to this exploration is the adoption of intersectional data practices—an essential strategy for designing technologies that empower women and marginalized communities rather than entrenching systemic biases.

(Continued)

(Continued)

The urgency of this dialogue cannot be overstated. Generative AI and other AI-powered systems act as force multipliers, capable of amplifying both opportunities and risks, perpetuating harmful stereotypes, privileging dominant voices, and marginalizing diverse perspectives— ultimately exacerbating systemic inequities, particularly for women and girls. However, when designed with intention and care and guided by intersectional principles, AI has the transformative potential to redefine inclusivity and equity, fostering a more just technological landscape. This work, therefore, advocates for the integration of human-centered frameworks and intersectional design principles as essential strategies for building systems that actively promote justice and equity at their core.

INTENDED AUDIENCE

Who is this book for? The audience spans policymakers, regulators, technology and design practitioners, researchers, academics, and engaged readers who see themselves as advocates for change. The insights and frameworks presented aim to shape policy, guide ethical technology development, advance interdisciplinary research, and enrich inclusive academic discourse. However, as a design leader with extensive industry experience, my emphasis naturally skews toward the design community, as reflected in the book's title and focus on design heuristics. The primary goal is to equip designers with tools and principles to navigate the complexities of creating equitable and user-centered technologies.

Policymakers and regulators play a critical role in enacting safeguards for women's health care data in the digital age. As I often assert, *nothing moves without policy*. However, much has moved without adequate protections, often to the detriment of women and girls. Design and technology practitioners are equally vital, as they directly shape the systems at the heart of these issues. This book provides actionable guidance to empower designers with ethical frameworks and technical insights, bridging user experience and systems design.

For example, following the guidance of Privacy by Design (Cavoukian, n.d.), secure hashing algorithms like Argon2 and bcrypt are essential technical measures to protect sensitive data. Yet, accountability also demands that these safeguards be communicated effectively through user interfaces that foster trust and transparency. Designers must ensure that protections are accessible and meaningful to everyday users, translating technical rigor into intuitive interactions.

Researchers and academics in fields such as data ethics, AI, women's health, and surveillance will also find this work relevant. By examining these intersections, the book highlights critical challenges and opportunities for collaboration. Meanwhile, engaged general readers are invited to see themselves not as passive observers but as advocates whose awareness can amplify the push for meaningful reform.

This work also includes and reflects a Canadian perspective, grounded in the Artificial Intelligence and Data Act, while drawing comparative insights from international frameworks such as the European Union's General Data Protection Regulation, the proposed U.S. Algorithmic Accountability Act, and the U.S. AI Bill of Rights. Additionally, the National Institute of Standards and Technology AI Risk Management Framework provides further context on managing AI's societal risks. However, these approaches often lack the practitioner's perspective on real-world implications—a gap this book seeks to address.

Ultimately, this heuristic supplement is a call to action for collaboration among all stakeholders to tackle the challenges posed by women's health care data and AI technologies, particularly in the post–*Roe v. Wade* landscape. By uniting efforts, we can create a digital future that prioritizes women's rights, dignity, and autonomy, ensuring that technology serves as a force for equity and justice.

DESIGN FOR EQUITY: A HEURISTIC FRAMEWORK FOR INCLUSIVE AND TRANSPARENT WOMEN'S HEALTH DATA SYSTEMS

The absence of robust AI regulatory frameworks and transparent oversight presents a dual-edged reality: unprecedented opportunities alongside profound ethical challenges, particularly for technology workers such as designers. Systemic biases, privacy violations, and the pervasive opacity of algorithmic systems increasingly threaten civil liberties, with marginalized communities disproportionately bearing the brunt of these harms (Eubanks, 2018; Noble, 2018). These dynamics underscore the urgent need for ethical, inclusive, and **human-centered design** principles to guide technological innovation.

Women's health technology represents a particularly critical intersection of these challenges. While advancements in this domain hold the promise of transforming health care delivery and empowering women, they simultaneously introduce unique risks. These include the potential misuse of sensitive reproductive data, the perpetuation of algorithmic bias in health care decision-making, and the exclusion of diverse needs during design processes (Criado Perez, 2019; Obermeyer et al., 2019). The safeguarding of women's health data requires not only technological innovation but also a rigorous and context-sensitive approach to addressing these ethical challenges with precision.

In such a complex environment, design heuristics emerge as a critical framework for fostering ethical and effective technology development. By embedding principles that prioritize human dignity, equity, agency, and accountability, heuristics enable technological systems to transcend narrow functional goals and align with broader societal values (Friedman & Hendry, 2019). This document proposes six foundational heuristics as a road map for designing systems that respect, protect, and empower women in the digital age (Lubin & Harris, 2024). These principles act as essential guardrails, promoting inclusivity, security, and transparency while addressing the pressing ethical concerns that characterize contemporary technological landscapes.

Each heuristic category highlights specific subheuristics that are critical to designing ethical and inclusive practices for managing women's health data. They are *user controls and affordances, accessibility, user consent, Privacy by Design,*[1] *human factors and ergonomics,* and *transparency.* The following section presents a comprehensive overview of the heuristic categories that serve as the foundational framework of this book, shaping its core principles and guiding its approach to human-centered design.

1. **User Controls and Affordances:** User controls and affordances focus on equipping women with robust mechanisms to manage and control their data-sharing and privacy settings. By emphasizing user-centric design, this heuristic enhances trust, respects individual autonomy, and supports ethical management of personal information.

2. **Accessibility:** Accessibility underscores the critical need for inclusivity within the context of data feminism. By adopting inclusive design principles for reproductive health platforms, this heuristic ensures equitable access for individuals with diverse abilities, enabling active participation and the dissemination of vital health care information and services.

3. **User Consent:** User consent highlights the foundational requirement for obtaining informed consent from women regarding the use of their reproductive health data. Through a transparent, characterization-driven approach to information design, this principle empowers women to make well-informed decisions about the sharing and management of their personal data.

4. **Privacy by Design:** Privacy by Design advocates for the proactive incorporation of privacy safeguards into the design and architecture of health technology. This principle is

---

1   Privacy by Design, established by Ann Cavoukian (n.d.), is a proactive framework integrating privacy into system architecture. It is founded on seven key principles: (1) proactive and preventive measures, (2) privacy as the default setting, (3) embedding privacy into design, (4) full functionality without trade-offs, (5) end-to-end security, (6) transparency, and (7) user-centric approaches. These principles ensure privacy protection while fostering trust, usability, and innovation.

essential to fostering **algorithmic trust**, ensuring the protection of sensitive health data, and upholding ethical standards in data practices.

5. **Human Factors and Ergonomics:** Human factors and ergonomics emphasize the importance of designing women's health technology systems that prioritize usability, comfort, and safety. By creating interfaces that are user-friendly and aligned with the cognitive and physical needs of women, this approach ensures that technology remains accessible, practical, and respectful of both privacy and personal safety.

6. **Transparency:** Transparency underscores the necessity of clarity and openness in the design of women's health care data systems. Transparent design practices are critical for building trust, as they ensure users have a clear understanding of how their data are collected, processed, and utilized, fostering confidence in the system's integrity.

ETHOTIC HEURISTICS AND THE CHARACTERIZATION OF WOMEN IN AI DESIGN

In this section, I expand on six key heuristic categories that inform the design of AI systems, exploring how the failure to adhere to these principles can influence the ways women are datafied, represented, and characterized within digital frameworks. Central to this discussion is the term *characterization*—a concept I draw from my academic background in computational rhetoric, particularly in the study of what I have termed, as previously noted, algorithmic ethopoeia (Lubin & Harris, 2024). Algorithmic ethopoeia is a rhetorical concept adapted from ancient Greek rhetoric, referring to how AI systems construct and portray the ethos, or character, of individuals they interact with. In classical rhetoric, ethopoeia was a technique used to craft a person's character through speech or writing; in the context of AI, it examines how algorithms generate, shape, and mediate digital representations of individuals, often influencing how they are perceived, categorized, and treated within technological systems. For everyday language, I will interchange with the more common language, computing the characterization of user.

In my research I argue that **characterization** of women by data systems is not a passive byproduct of AI design but an active and consequential process that must be consciously addressed. AI's capacity to shape and reflect women's identities—whether through data-driven representations, predictive models, or user profiles—raises critical questions about inclusivity and fairness. Missteps in this area risk alienating the very users these systems aim to serve. In contrast, embedding ethotic heuristics into AI design offers a pathway toward more empathetic, accountable, and representative technologies (Lubin & Harris, 2024).

In this context, characterization goes beyond mere surface-level representation, understanding the complex realities and contexts of women's lived experiences, perspectives, and needs. AI systems, reliant on data-driven processes, frequently construct characterizations that lack the depth and nuance of human identity, instead reducing individuals to fragmented data points. This oversimplification and lack of understanding of the context can perpetuate biases and reinforce narrow stereotypes—one day an abortion is legal, and the next day it is no longer so, as we have seen in the shocking reversal of *Roe. v Wade*. To counteract these limitations, thoughtful design strategies must be employed— ones that move beyond reductive portrayals and actively incorporate diverse narratives, ensuring that AI systems reflect the richness and complexity of human identity rather than distilling it into algorithmic shorthand.

The following proposed heuristics guide this effort:

1. Identifying and addressing common technological pain points experienced by women

2. Prioritizing design elements that reflect women's unique health journeys and life stages

3. Mitigating stereotypes and biases that can shape AI-driven characterizations

4. Incorporating diverse female voices throughout the development process

5. Acknowledging the intersection of gender with race, class, culture, and other social factors

6. Designing adaptable, flexible AI systems that reflect the varied needs of women

By embedding ethotic heuristics—a framework that considers how AI-driven systems construct and represent human character—into AI design, we can mitigate the potential harms these systems may pose to women. When AI relies on reductive or biased data, it risks misrepresenting women, reinforcing stereotypes, and overlooking their complexity. Integrating ethotic heuristics ensures that AI technologies do more than just process data—they engage with women as whole individuals, acknowledging their diverse identities and lived experiences. This approach fosters trust, promotes equity, and establishes a stronger ethical foundation for AI development.

The Human Cost of Datafication: Resisting Reductionism in AI Design and Oversight

AI-driven systems increasingly influence how we are represented, understood, and categorized. Yet, these systems often reduce human complexity to fragmented data points, failing to capture the full depth of individual identities. As I explored in my recent blog post, "Wake Up, Humans! Our Data Crisis Is Really a Humanity Crisis—From Bias to Greed: The Many Data Exploits" (Lubin, 2024), this crisis extends beyond data—it is a crisis of humanity itself.

This issue is particularly urgent in health care, where AI's role in data collection and processing reveals both its potential and its dangers. While concerns about how women and girls are represented in AI systems are critical, it is equally important to examine the broader political and economic forces shaping these technologies. As noted in my earlier research (Lubin & Fan, 2025; Lubin & Harris, 2024), the intersection of state power and Big Tech capitalism raises profound ethical concerns, particularly regarding access to personal data and the elusiveness of true user consent. Zuboff (2019) describes how tech corporations engage in "surveillance capitalism," where user data are harvested to predict and influence behavior, often without explicit consent, creating asymmetrical power dynamics between individuals and corporations. Similarly, Floridi et al. (2019)

argue that Big Tech wields influence comparable to nation-states, exacerbating concerns about digital governance and regulatory capture. Couldry and Mejias (2019) extend this argument, framing the appropriation of personal data as a form of digital colonialism that disproportionately impacts marginalized communities. The global implications of these dynamics are further explored by Taylor and Broeders (2015), who highlight how the datafication of the Global South reinforces socioeconomic inequalities and consolidates Big Tech's dominance over information infrastructures. Milan and Treré (2020) add to this discussion by illustrating how vulnerable populations, often characterized as the "data poor," are subjected to heightened forms of digital surveillance without meaningful consent, particularly in times of crisis, such as the COVID-19 pandemic. These works collectively emphasize the urgency of addressing the ethical dilemmas at the nexus of state control and corporate data monopolies, advocating for stronger regulatory frameworks and data sovereignty measures.

Even when designed with empathy, AI systems operate within frameworks that favor measurable outcomes over lived experiences. Human oversight is necessary but insufficient, as it often reinforces the same reductionist logic that governs these technologies. True accountability demands a fundamental shift—from mere regulatory compliance to a deeper ethical reckoning with the societal impact of AI.

To move forward, we must advocate for inclusive and **empathetic design** practices while also challenging the larger systems that commodify human data. This dual approach—ensuring fair representation in AI while critically interrogating the economic and political structures that shape these technologies—is essential. The heuristic guidance in the following chapters offers a road map for achieving this balance. However, our responsibility extends beyond improving representation; we must resist the normalization of datafication as the default mode of human interaction. By prioritizing human dignity over mere functionality, we can shape a future where technology serves people—not the other way around.

KEY ACTIONS AND PRINCIPLES FOR
ETHICAL TECHNOLOGY DESIGN:
PROTECTING WOMEN'S HEALTH CARE DATA

From an experiential perspective, with nearly 20 years as a design research leader—much of that time spent guiding teams in handling user data to inform design decisions—I find myself deeply alarmed by the current state of women's health care data. These data, often sensitive and deeply personal, demand far greater safeguards to prevent misuse or exploitation. My awareness of these risks was heightened during my time at Siemens Research, where I worked on medical and health care applications. There, it was essential to undergo Health Insurance Portability and Accountability Act training, reinforcing the critical importance of privacy protections in health care technology. That experience underscored for me that ethical data handling is not just a regulatory requirement but a moral imperative. Today, as digital health technologies rapidly evolve, the need for transparency, accountability, and an unwavering respect for privacy has never been more urgent. Without these principles at the core of system design, we risk reinforcing inequities and eroding the very trust that health care systems depend upon.

These principles are not just safeguards for women's rights— they are the foundation for cultivating trust in the very systems designed to support their health care needs. Without a commitment to ethical, transparent, and accountable design, these technologies risk reinforcing systemic inequities rather than alleviating them. To build meaningful protection and ensure that women's health data are handled with the care they deserve, we must adopt a reimagined approach to design and implementation—one that actively integrates privacy, agency, and inclusivity at every stage. Achieving this requires ethical collaboration across several critical areas, each of which plays a pivotal role in shaping a future where technology genuinely serves women rather than exploiting them:

1. **Risk-Free Auditing and Accountability:**

   These systems must operate transparently, upholding trust and ensuring compliance with laws such as General Data Protection Regulation and U.S. data protection standards, including state-level acts like the California Consumer Privacy Act.

2. **Adopting Guiding Heuristics for Responsible AI Design:**

Ethical AI systems, especially those managing women's health care data, must align with key principles of design, broadly outlined in these categories:

i. **User Controls and Affordances:** Empower women and girls with control over their data, offering tools to decide how they are accessed, managed, and utilized in a way that aligns with their needs and values.

ii. **Accessibility:** Build systems that are inclusive and accessible to all, particularly women and girls, accommodating different abilities, socioeconomic conditions, and cultural contexts.

iii. **User Consent:** Provide mechanisms for clear, informed, and revocable consent, ensuring that women and girls fully understand and control how their data are collected, shared, and used.

iv. **Privacy by Design:** Enforce robust data protection measures and prioritize anonymity to safeguard sensitive information about women and girls, particularly in vulnerable contexts.

v. **Human Factors and Ergonomics:** Design systems that reflect the diverse needs, experiences, and perspectives of women and girls, ensuring that their unique requirements are prioritized and addressed.

vi. **Transparency:** Clearly communicate how AI systems operate, how data are handled, and the associated risks, ensuring women and girls can make informed decisions about their engagement with technology.

The implementation of these guiding principles extends far beyond the immediate objective of safeguarding women's health care data—it is a pivotal step toward fostering a broader culture of justice, equity, and ethical responsibility in technology. By embedding these heuristics into system design, we create digital environments that not only protect individual rights but also proactively minimize risks and reinforce trust. In doing so, we set a higher standard for technological development—one that prioritizes human dignity, transparency, and fairness, ensuring that digital health solutions truly empower rather than exploit.

READER DECLARATION OF COMMITMENT

As technology, power, and social justice become increasingly intertwined, it is essential to recognize our role in shaping digital futures. This Reader Declaration serves as an invitation to engage critically, reflect deeply, and commit to ethical considerations in discussions of AI, privacy, and systemic inequities. By acknowledging the responsibilities that come with knowledge, we collectively affirm the urgency of accountability, advocacy, and transformative action in the digital age.

As a reader of *Design Heuristics for Emerging Technologies: AI, Data, and Human-Centered Futures: Considerations for the Rights of Women*, I acknowledge the profound influence of emerging technologies on society and the responsibility to shape their development in ways that uphold equity, inclusivity, and justice.

By signing this declaration, I commit to the following:

1. **Advancing Equity:** Prioritizing the rights and representation of women and marginalized communities in the design, development, and deployment of emerging technologies

2. **Ethical Design Practices:** Applying human-centered and inclusive design principles to ensure technologies align with societal values and do not perpetuate bias or discrimination

3. **Transparency and Accountability:** Advocating for systems that are transparent in their operation, explainable in their decisions, and accountable for their outcomes

4. **Empowerment via Collaboration:** Engaging diverse voices, particularly women, in cocreating solutions that address global challenges and improve quality of life

5. **Continuous and Intentional Reflection:** Staying vigilant and responsive to the evolving impacts of technology, fostering a culture of learning and adaptation in the pursuit of ethical innovation

My Commitment to Ethical Action

I, _____, understand the importance of embedding these principles into my professional and personal practice. By signing this declaration, I affirm my dedication to fostering a future where technology empowers and uplifts all individuals, particularly those historically underrepresented or disadvantaged.

Date: _____

Signature: _____

The importance of aligning technological advancements with human rights and dignity has been emphasized in global policy discussions. For example, UNESCO's *Recommendation on the Ethics of Artificial Intelligence* (n.d.) underscores the necessity of ensuring AI development is guided by principles of fairness, accountability, and respect for human rights. This framework advocates for policies that mitigate algorithmic biases and prevent the reinforcement of existing societal inequities, reinforcing the urgency of ethical AI governance. The future we create begins with the actions we commit to today. Together, we can design a world where technology serves the rights and dignity of all people.

Reader Note (Reflection):

_____

_____

_____

_____

_____

_____

_____

ABOUT THE CASE STUDIES
USED IN THIS BOOK

The growing prevalence of smartphone applications tailored for pregnant individuals reflects a significant shift in how women and girls now access and engage with health information. Studies indicate that a significant proportion of pregnant women utilize pregnancy-related apps to support their health and track their pregnancy progress. For instance, a 2021 study found that over 50% of pregnant women use such apps, browsing through thousands available on app stores (Frid et al., 2021). Another study reported that 72% of survey participants used pregnancy apps during their pregnancy (Lazarevic et al., 2023). Additionally, during the COVID-19 pandemic, approximately 77.9% of respondents reported using pregnancy-related mobile apps, highlighting an increased reliance on digital health technologies during that period (Grand View Research, 2023). Despite their widespread use, many apps face criticism for lacking credibility, accuracy, and evidence-based clinical guidance, with some even incorporating potentially harmful functionalities (Asadollahi et al., 2025; Jones & Taylor, 2018).

Prior research on pregnancy apps has largely focused on their content reliability, expert involvement, and commercialization techniques, revealing substantial gaps in their design and execution. For instance, studies by Chen et al. (2019) and Cao et al. (2024) highlight the prevalence of unregulated health advice and misinformation, underscoring the need for greater oversight in app development. Furthermore, user perception studies suggest that commercialization techniques, such as in-app advertising and affiliate marketing, detract from app usability and diminish user trust (Lupton & Pedersen, 2016).

Emerging concerns also extend to app functionalities, with previous analyses noting a tendency toward "feature creep"—the inclusion of numerous features that may compromise app quality and usability. Functionalities such as fetal kick counters, while beneficial, are often poorly integrated and lack clinical validation (Kumar & Singh, 2020). Moreover, only a fraction of pregnancy apps explicitly mention the involvement of medical experts, a critical factor in establishing their credibility (BioMed Central, 2025).

Despite these limitations, pregnancy apps holds promise as tools for enhancing maternal health when developed using evidence-based frameworks. Financial incentives, such as government or institutional grants, could encourage the production of high-quality, clinically sound applications. Additionally, health care providers play a crucial role in bridging the gap between app developers and end users, guiding pregnant individuals toward trustworthy platforms and helping mitigate risks associated with unreliable apps.

This review underscores the pressing need for collaborative efforts among developers, health care professionals, and regulatory bodies to establish standards for pregnancy app development. Doing so could enhance the apps' reliability, address user concerns, and ultimately improve maternal health outcomes.

Health care serves as the central focus of this book and my academic research due to its critical importance as a life-or-death sector. The consequences of biased AI systems in health care disproportionately harm women, particularly those from marginalized communities, exacerbating inequities in access, treatment, and outcomes. By prioritizing health care, I aim to highlight these disparities and propose practical heuristics for developing equitable AI systems.

This focus underscores the urgent need to reform data governance and design practices to ensure health care data become a tool for equity rather than exclusion. Through this lens, I seek to contribute actionable insights for creating technologies that prioritize fairness and justice, especially in such a vital domain.

The case studies (e.g., Flo, Clue, and Glow, among others) that informed the development of these heuristics also offer crucial contextual foundations and catalytic insights for their creation. Moreover, they serve as concrete illustrations of the heuristics' relevance and necessity. While certain instances of poorly designed systems have been mitigated through public backlash or legal interventions, referencing these cases remains imperative. They illuminate recurring patterns of harm and systemic oversight, underscoring the persistent challenges that necessitate robust ethical frameworks in technology design and implementation.

Consequently, the heuristics developed here are tailored to this context, addressing key issues such as user privacy, agency, and the intersections of gender and health data. By focusing on health care, these heuristics tackle critical issues like the exploitation of intimate health data, the reinforcement of gender biases in algorithmic systems, and the ethical challenges posed by invasive tracking technologies. However, the underlying principles—rooted in equity, inclusivity, and human dignity—are broadly applicable and can inform ethical design across diverse domains, adapting to the specific dynamics of each new context.

If the research had instead focused on a different specialized group—such as female pilots—the heuristics would naturally adapt to address context-specific challenges, such as anthropometric diversity in cockpit design or reducing gender bias in training simulations and performance evaluations. This same principle was central to the women's health movement, which emerged in the 1960s and 1970s (see Table I.1) to challenge medical research that systematically excluded women and failed to account for gender-specific health needs (Boston Women's Health Book Collective, 1973). Just as activists fought for medical studies and health care policies that reflected women's lived experiences, heuristic frameworks in technology must be designed with the same responsiveness to ensure equitable and inclusive outcomes. This contextual adaptability underscores that heuristics are not one-size-fits-all solutions; they must be responsive and informed by the lived realities of the groups they aim to support. The goal is to ensure that design practices are both theoretically robust and practically effective in meeting the unique needs of real-world applications.

1960s	The women's health movement begins, with increasing activism for women's reproductive rights and health care.
1973	The U.S. Supreme Court's decision in *Roe v. Wade* legalizes abortion, a key victory for women's health rights.
1983	The Public Health Service Task Force on Women's Health Issues is established to address women's health concerns.
1986	The National Institutes of Health enact the Inclusion of Women and Minorities in Clinical Research policy, mandating equal participation of women and minorities in medical studies.
1990	The Women's Health Protection Act (WHPA) is passed, leading to the creation of the Office of Research on Women's Health.
1993	Congress passes the NIH Revitalization Act, increasing funding and research efforts focused on women's health.

1994	The International Conference on Population and Development, held in Cairo, marks a paradigm shift by emphasizing reproductive health and rights, moving away from demographic targets to focus on individual needs.
1994	The Violence Against Women Act is enacted in the United States, which provided comprehensive measures to combat domestic violence, sexual assault, and stalking, significantly enhancing legal protections for women.
1998	The Society for Women's Health Research advocates for women's inclusion in clinical trials, leading to increased awareness and policy changes ensuring women's representation in medical research, addressing historical gender disparities.
2006	Approval of the human papillomavirus (HPV) vaccine represented a significant advancement in preventing cervical cancer and other HPV-related diseases among women.
2017	Millions worldwide participate in the Women's March, advocating for women's rights, including health care access, reproductive rights, and addressing violence against women.
2022	The U.S. Supreme Court's decision in *Dobbs v. Jackson Women's Health Organization* overturns the landmark 1973 *Roe v. Wade* ruling, ending federal protection for abortion rights and shifting regulatory power to individual states.
2023	In response to changing abortion laws, the WHPA is reintroduced in Congress to protect the right to access abortion services nationwide, aiming to codify reproductive rights into federal law.

Table I.1   Women's Health Movement, 1960–2023

Sources: Adapted from *Blog: Women's Health Movement: A Brief History*, by B. Azevedo, T. Juarez, & A. Taylor, March 8, 2023, Weitzman Institute (https://www.weitzmaninstitute.org/blog-womens-health-movement-a-brief-history/); *Report of the International Conference on Population and Development, Cairo, 5–13 September 1994*, by United Nations, 1995 (https://www.un.org/development/desa/pd/sites/www.un.org.development.desa.pd/files/icpd_en.pdf); Violence Against Women Act of 1994 (https://www.congress.gov/bill/103rd-congress/house-bill/3355); *A Timeline to Making Women's Health Mainstream*, by Society for Women's Health Research, 2025 (https://swhr.org/about/1977-1989-timeline/); *Human Papillomavirus (HPV) Vaccine Safety*, by U.S. Food and Drug Administration, March 6, 2025 (https://www.cdc.gov/vaccine-safety/vaccines/hpv.html); *Our Vision*, by Women's March, 2025 (https://www.womensmarch.com/about-us); *Dobbs v. Jackson Women's Health Organization*, 2022 (https://www.oyez.org/cases/2021/19-1392); Women's Health Protection Act of 2023 (https://www.congress.gov/bill/118th-congress/house-bill/12).

CHAPTER 1

User Controls and Affordances

Empowering User Agency Through User-Centric Data Sharing and Privacy Settings

User agency is a pivotal aspect of ethical technological design, especially when it comes to data sharing and privacy settings. This heuristic underscores the importance of empowering users to have meaningful control and autonomy over their personal information. Through a characterization-driven approach, designers can create intuitive and adaptable interfaces that align with individual user preferences, fostering trust, satisfaction, and ethical data management. By prioritizing user-centric design, technology shifts from dictating user behavior to enabling individuals to take charge of their data and digital experiences.

THE HUMAN COST OF DATAFICATION: RESISTING THE REDUCTION OF WOMEN AND HUMANITY TO DATA POINTS

Our increasing reliance on data-driven technologies exposes a growing concern regarding the fragmentation of identity and representation. In my early 2024 blog post, "Wake Up, Humans! Our Data Crisis Is Really a Humanity Crisis—From Bias to Greed: The Many Data Exploits," I explore how the current exploitation of data reflects incomplete and often misleading glimpses of individual identities, obscuring the complexity of human experience (Lubin, 2024). This crisis extends beyond the realm of data—it underscores a broader existential concern rooted in the erosion of holistic representations of humanity. Nowhere is this phenomenon more apparent than in the health care sector, where data collection and analysis present both significant opportunities and profound risks. The ways in which personal data are harnessed, commodified, and interpreted reveal not only technological advancements but also the perils of reducing people to fragmented datasets. Amid this changing tech and sociopolitical landscape, the promise of personalized medicine and health care is tempered by the risk of reducing our patient narratives to simplistic data points that overlook the richness of our individual experiences. Given this, this tension between innovation and vulnerability has sparked urgent debates around consent, privacy, and the ethical stewardship of our personal health information. As health care continues to traverse new digital frontiers, establishing robust safeguards is imperative to ensure that technological progress enhances—not diminishes—the human essence at its core. And nowhere is it as critical than in the characterization of female health care datafication, informed by relentless **capitalistic reductionism.**

CAPITALISTIC REALISM AND THE
REDUCTION OF HUMAN COMPLEXITY

While the mischaracterization of women and girls by artificial intelligence (AI) systems warrants immediate attention, it is imperative to situate this issue within the larger socioeconomic and technological landscape. AI operates within a framework of **capitalistic realism,** a paradigm that views human experience as a series of quantifiable data points, optimized for efficiency and scalability. This reductionist logic stands in stark opposition to the inherent complexity, unpredictability, and emotional richness that characterize our human existence. The prioritization of data-driven outcomes reflects a systemic inclination to flatten nuanced identities into simplified metrics, diminishing the full spectrum of lived experiences. We must actively challenge this reductionist stance by reimagining design systems that not only optimize efficiency but also embrace the richness of our human experiences. By centering empathy at every stage of the design process, we ensure that technological innovation remains grounded in the unpredictable, emotional, and multifaceted nature of our existence, ultimately creating systems that honor and amplify the true complexity of what it means to be human— and, in context, what it means to be female.

THE LIMITS OF EMPATHY
IN SYSTEM DESIGN

Even AI systems designed with empathy are constrained by the overarching logic of quantification and categorization. Despite efforts to incorporate inclusive design principles, the structural limitations of these systems often privilege measurable outputs over qualitative experiences. This underscores the necessity for **human oversight** in AI decision-making processes. However, oversight alone is insufficient. The individuals responsible for monitoring these systems frequently operate within the same

reductionist framework, reinforcing existing biases and perpetuating the detachment between data and authentic representation.
Achieving meaningful accountability requires moving beyond
procedural oversight to engage with the deeper ethical questions
surrounding AI and data usage. Moreover, establishing robust,
multilevel accountability mechanisms that invite diverse and
representational perspectives is crucial to challenging entrenched
assumptions about data design practices. This critique sets the
stage for a deeper examination of system design—one that inter-
rogates not just the visible outputs but the very foundations
upon which these technologies are built.

DUAL IMPERATIVE: REPRESENTATION
AND SYSTEMIC CRITIQUE

Addressing the ethical implications of AI involves a dual focus—
advocating for the inclusive representation of marginalized
groups while simultaneously challenging the systemic forces that
commodify human identity. From a **systems design** and user
interface design perspective, this approach calls for the development of technologies that acknowledge and respect the
multifaceted nature of human experience. Inclusive design must
extend beyond surface-level representation to interrogate the
economic and ideological structures that drive technological
development. This critical engagement is essential to resist the
normalization of **datafication**—the pervasive practice of reducing individuals to mere data objects—and to promote a vision of
technology that prioritizes humanity over functionality.

HEURISTIC GUIDANCE FOR
INCLUSIVE AI DESIGN

The following chapters offer heuristic strategies for designing AI
systems that reflect and respect the diverse experiences of
women and girls. These heuristics emphasize not only the importance of representation but also the necessity of interrogating the
socioeconomic frameworks that shape technological innovation.
Key considerations include the following:

- **Challenging data-centric paradigms** by designing AI that reflects holistic identities rather than fragmented data points

- **Resisting datafication** by advocating for systems that prioritize user agency, adaptability, and emotional depth

- **Integrating diverse voices** throughout the design process to ensure equitable representation and avoid reinforcing harmful stereotypes

By embracing these principles, we can contribute to the development of AI-powered data design systems that not only improve the visibility and characterization of women and girls but also foster a broader cultural shift toward ethical and human-centered technology. This process demands an ongoing commitment to reevaluating the paradigms that govern what is often seen as technological advancement, ensuring that emerging systems reflect the richness and diversity of the human experience.

A social networking platform dedicated to women's health provides a practical example of user controls and affordances. BabyCenter (www.babycenter.com), a popular community for pregnancy and parenting discussions, has faced criticism for limited user controls. Users reported challenges (see BabyCenter, 2021) in managing their privacy settings, such as difficulty removing sensitive posts or controlling who could view their personal data. This lack of affordances not only compromised privacy but also discouraged meaningful participation, especially in sensitive discussions about reproductive health.

In contrast, consider a platform like Peanut (www.peanut-app.io), which connects women to discuss fertility, pregnancy, and parenting. Peanut emphasizes **user-centered privacy design** by providing intuitive controls for privacy and content sharing. The app allows users to do the following:

- Adjust visibility settings to decide whether their profile is public or private, and select who can interact with them.

- Easily delete posts or comments that they no longer wish to share.

- Customize notifications to avoid overwhelming or intrusive alerts.

- Control data sharing preferences through a straightforward privacy dashboard, empowering users to opt in or out of data usage for research or marketing purposes.

By embedding these affordances directly into the platform's design, Peanut fosters a safe and supportive environment where women feel in control of their participation. This proactive approach not only protects users but also encourages trust, enabling open and meaningful engagement around sensitive health topics. Platforms that adopt similar user controls can create digital spaces that are not only functional but also empowering.

In such a platform, design choices might include the following:

- **Customizable Privacy Settings:** Allowing users to regulate who can access their health-related posts or personal information

- **Clear Consent Mechanisms:** Prompting users for explicit consent whenever sensitive data are requested

- **Granular Data-Sharing Options:** Enabling users to share specific data only with selected groups or individuals

This approach fosters a safe and empowering digital environment where women can confidently engage and share.

Proposed Heuristics for User Controls and Affordances in System Design That Considers Women's Rights and Equity

Designing AI systems and digital platforms that respect women's rights and promote equity requires intentionality at every stage of development. User controls and affordances—features that empower individuals to shape their interactions with technology—serve as critical touchpoints for fostering agency, safety, and inclusivity. However, without conscious design interventions, these systems risk perpetuating biases, reinforcing harmful stereotypes, and excluding marginalized voices.

The following heuristics provide a framework for embedding equity-driven user controls and affordances into system design. By prioritizing transparency, adaptability, and intersectional representation, these guidelines aim to ensure that technological environments not only reflect but actively support the diverse needs and lived experiences of women and girls.

1.1. Privacy Settings Customization

    1.1.1. Allow users to tailor privacy settings to suit their comfort levels, specifying who can view their profiles, access their data, or interact with them.

1.2. Data-Sharing Consent

    1.2.1. Require explicit consent for accessing or sharing sensitive data. Present the information in clear, jargon-free language, with opt-out options readily available.

1.3. In-App Education

    1.3.1. Provide resources to educate users about privacy settings, explaining their implications and benefits.

    1.3.2. Use tutorials, FAQs, or interactive guides to promote informed decision-making.

1.4. User-Friendly Interface

    1.4.1. Design interfaces to be intuitive and easily navigable:

- Employ logical layouts and visual cues to help users locate and manage privacy settings.

1.5. Granular Control

    1.5.1. Offer granular options for data sharing:

- For example, users could choose to share menstrual health data only with health care providers or specific support groups.

1.6. Opt-Out Mechanisms

    1.6.1. Allow users to opt out of sharing certain types of data or participating in specific platform features entirely, ensuring they remain in control of their engagement.

1.7. Access to Data

    1.7.1. Provide users with easy access to their data, allowing them to review, edit, or delete information as they see fit.

1.8. Ethical Design Choices

1.8.1. Adopt an ethical lens in design decisions, ensuring that

- Users retain autonomy over their data.

- Platform practices are transparent, respectful, and aligned with users' expectations.

Prioritizing user controls and affordances in design reflects a commitment to empowering women, building trust, and enhancing satisfaction. By adopting a characterization-driven approach, designers can create systems that thoughtfully consider the diverse needs and preferences of women, enabling meaningful control over their data and interactions. This approach fosters an environment where women feel respected, secure, and valued, addressing unique concerns such as privacy, safety, and inclusivity.

Empowering women with agency in data design practices transforms technology from a passive tool into an active enabler of autonomy and confidence, bridging the gap toward more ethical and equitable technological practices. By embedding these principles, designers not only support women's individuality but also contribute to shaping a technological future that champions respect, equity, and empowerment.

CHAPTER 2

Accessibility

Ensuring Inclusion in Data Feminism Through Reproductive Health Platforms

Accessibility serves as a foundational pillar of ethical technology design, particularly within the framework of data feminism. This category emphasizes the critical need for inclusive platforms in reproductive health, addressing physiological and cognitive atypicalities related to mobility, vision, hearing, and reproductive health.[1] By integrating assistive technologies and adhering to inclusive design principles, these platforms can empower individuals of all abilities, ensuring equitable access to vital information and services. This guidance provides insights, key heuristics, and real-world examples to inform the development of accessible reproductive health technologies. The subheuristics within this category are focused on established classifications of accessibility needs, offering targeted strategies to accommodate atypicalities and enhance inclusivity.

Accessibility ensures digital platforms are designed for all users, including those with physical, sensory, or cognitive challenges, enabling equitable navigation and engagement. Data feminism also highlights the need to address inequities affecting marginalized groups, advocating for technologies that uplift diverse voices and ensure inclusive access to resources.

THE POWER OF INCLUSIVE DESIGN

Inclusive design is more than just a best practice—it is a fundamental approach to ensuring equitable access to technology and health care. By prioritizing accessibility, inclusivity fosters autonomy, expands vital resources, and reinforces ethical standards in digital innovation:

---

1   Atypicalities refer to variations in reproductive health conditions or experiences that deviate from the commonly understood norms but are still significant and valid. For example, individuals with uterine anomalies, such as a bicornuate uterus, may face unique reproductive health challenges that require specific guidance and support. A bicornuate uterus is a congenital uterine anomaly resulting from incomplete fusion of the Müllerian ducts, often leading to reproductive challenges such as recurrent pregnancy loss, preterm labor, and fetal malpresentation (Kaur & Panneerselvam, 2023). Management strategies may include a procedure known as a metroplasty to improve uterine shape and high-risk pregnancy monitoring to mitigate complications.

- **Enhancing Autonomy:** Accessible platforms promote independence and confidence.

- **Broadening Access:** Inclusive reproductive health tools provide vital resources for users with varying abilities.

- **Advancing Ethical Tech:** Inclusive design upholds dignity and combats systemic digital inequities.

Consider an online platform designed to provide reproductive health resources, such as menstrual tracking, fertility planning, and access to health care providers. Failure to address accessibility in such a platform can have serious consequences, both for users and for the credibility of the platform itself. In recent years, several women's health apps have faced criticism for not adhering to accessibility guidelines, thereby limiting their usability for women with disabilities. A study published in *BMC Pregnancy and Childbirth* found that over 50% of pregnant women use smartphone pregnancy apps, yet many of these platforms lack credibility, provide inaccurate health information, and fail to follow evidence-based clinical guidelines (Nissen et al., 2024). These shortcomings not only create barriers for users with disabilities but also contribute to misinformation and potential health risks, underscoring the urgent need for more inclusive and rigorously evaluated digital health tools.

In another notable case, Flo, a widely used menstrual tracking app, faced backlash not only for data privacy concerns but also for failing to meet accessibility standards, such as intuitive navigation for users with cognitive impairments (Federal Trade Commission, 2021a, 2021b).

The lack of inclusivity reinforced barriers for individuals already underserved by mainstream health care systems, underscoring the broader systemic neglect in digital health design. By contrast, an accessible platform would prioritize features like compatibility with assistive technologies, including screen readers and voice control interfaces, as well as culturally and linguistically diverse content. Such a platform would demonstrate a commitment to inclusivity, ensuring that reproductive health tools empower all users, particularly those from marginalized communities.

This oversight by Flo excluded a significant segment of women users from accessing vital reproductive health tools, despite the app being marketed as an inclusive solution for fertility planning.

Such inaccessibility not only alienates some users but also perpetuates inequities by denying individuals with disabilities access to critical health resources.

Accessibility is not merely a technical add-on; it is a moral and practical imperative in the creation of equitable and effective health technologies. An inclusive approach would include the following:

- **Assistive Technologies:** Integrating tools such as screen readers, text-to-speech features, and sign language interpretation
- **Cultural Sensitivity:** Offering resources in multiple languages and considering cultural nuances in health-related communication
- **Intuitive Navigation:** Ensuring the platform is simple to use, with clear instructions and logically organized content

Such a platform would reflect a commitment to accessibility by addressing diverse needs while maintaining the sensitivity required in reproductive health contexts.

Proposed Heuristics for Accessible Design in System Design That Consider Women's Rights and Equity

Accessibility in system design extends beyond compliance; it is a known pathway to equity, ensuring that technology serves diverse populations, including women and marginalized communities. Designing accessible systems through the lens of women's rights requires addressing not only physical and cognitive barriers but also social, economic, and cultural dimensions of exclusion. For instance, the universal design approach in technology advocates for inclusivity, enhancing independence, job opportunities, social connectivity, safety, education, and economic empowerment for women with disabilities (Centre for Excellence in Universal Design, 2025). Cocreation that elicits feedback from users, in context of use, improves the user experience, while legal frameworks push for equitable access. This approach reduces economic barriers, fosters diversity, and promotes collective societal growth, again making the proposed heuristics so important.

The following heuristics, accordingly, outline strategies for embedding accessibility into the core of system design, inclusive of user interface and other elements. By prioritizing inclusivity, intersectionality, and user empowerment, these guidelines seek to create technologies that recognize and respond to the varied needs and experiences of women, fostering environments where all users can engage fully and equitably.

1.9.   Inclusive Design

1.9.1.   Use clear, concise, and inclusive language to enhance comprehension, and avoid jargon or culturally specific terms that may alienate some users.

1.9.2.   Organize information logically and hierarchically to reduce cognitive load, improve navigation, and enable users to quickly locate essential information.

1.9.3.   Incorporate consistent design patterns and visual cues to foster familiarity and predictability, reducing the learning curve for new users.

1.9.4.   Provide multiple ways to access information, such as text alternatives for visual elements, audio descriptions, and screen reader compatibility, to accommodate diverse preferences and abilities.

1.9.5.   Ensure flexibility in interaction methods, such as offering voice commands, keyboard navigation, and touch inputs, to suit various user needs.

1.9.6.   Use visual and functional simplicity by minimizing clutter, grouping related elements, and avoiding unnecessary distractions.

1.9.7.   Include error prevention and recovery mechanisms, such as clear instructions, undo options, and context-sensitive help, to build user confidence and reduce frustration.

1.9.8.   Offer customization options to adapt the interface to individual preferences, such as adjustable font sizes, themes, or layouts that enhance accessibility.

1.9.9.   Conduct usability testing with diverse user groups, especially women and girls with varying cognitive and physical abilities, to identify potential barriers and ensure inclusivity.

1.9.10.   Prioritize real-time feedback for user actions to provide clear acknowledgment and guidance, ensuring that users feel in control of the system.

1.10. Assistive Technology Integration

Incorporate features that accommodate users with diverse abilities:

1.10.1.   Text-to-Speech Functionality: Enable users with visual impairments to access information by providing high-quality, customizable text-to-speech features that support multiple languages and accents.

1.10.2.   Sign Language Interpretation: Offer sign language options, including video overlays or avatars, to enhance accessibility for individuals who are hard of hearing or deaf. Ensure support for regional and international sign languages.

1.10.3.   Keyboard Navigation Support: Design interfaces that allow users with mobility impairments to navigate efficiently using keyboard-only inputs, with clear focus indicators and logical tab order.

1.10.4.   Speech-to-Text Capability: Provide tools that allow users with limited mobility or dexterity to input text or commands using voice recognition technology, ensuring high accuracy and multilingual support.

1.10.5.   Alternative Input Devices: Ensure compatibility with assistive input technologies, such as switch controls, eye-tracking systems, and head pointers, to support users with severe physical disabilities.

1.10.6.   Customizable Interfaces: Allow users to personalize settings like font size, color contrast, and screen layout to meet their individual needs and preferences.

1.10.7.   Real-Time Captioning: Include live or automated captioning for audio content, such as videos and meetings, to support users with hearing impairments or auditory processing challenges.

1.10.8.   Vibration or Tactile Feedback: Use haptic feedback to provide additional sensory cues for users with hearing or vision impairments.

1.10.9.   Compatibility With Screen Readers: Ensure full integration with screen reader technology, using accessible HTML, ARIA (Accessible Rich Internet Applications) roles, and semantic markup to guide navigation.

1.10.10.   Testing With Assistive Technologies: Regularly test the platform with real assistive devices and software used by diverse populations to identify and resolve compatibility issues.

1.11. Reproductive Health Considerations

Ensure the platform provides comprehensive and inclusive reproductive health resources:

1.11.1.   Address a Wide Range of Health Conditions and Atypicalities: Include information and support for various reproductive health conditions, such as polycystic ovary syndrome, endometriosis, menopause, fertility challenges, and pregnancy complications.

1.11.2.   Offer Tailored Advice and Information: Provide personalized content based on factors such as age, cultural background, gender identity, sexual orientation, and specific health concerns, ensuring relevance for diverse populations.

1.11.3.   Incorporate Menstrual Health Tracking and Education: Include tools for tracking menstrual cycles, symptoms, and ovulation while providing educational resources on menstrual health and hygiene.

1.11.4.   Ensure Privacy and Confidentiality: Design features to protect sensitive health data, such as encrypted storage and user-controlled sharing options, to build trust and ensure user safety.

1.11.5.   Provide Multilingual and Culturally Sensitive Resources: Offer content in multiple languages and ensure that resources respect cultural beliefs and practices related to reproductive health.

1.11.6.   Include Accessibility Features: Ensure reproductive health content is accessible to users with disabilities, such as screen reader compatibility, text-to-speech options, and visual adjustments.

1.11.7.   Offer Mental Health Support Related to Reproductive Health: Address mental health issues like postpartum depression, anxiety around fertility challenges, or menopause-related mood changes by providing integrated support and resources.

1.11.8.   Facilitate Connection to Health Care Providers: Include features for locating, contacting, or scheduling appointments with local reproductive health specialists, clinics, or support organizations.

1.11.9.   Regularly Update Content: Ensure that health advice and resources remain current by incorporating the latest medical research, guidelines, and best practices.

1.12. Sensitive and Empathetic Design

Standard design processes, geared toward a female population, must incorporate characterization to understand women's experiences, societal contexts, and preferences:

1.12.1.   Engage Directly With Diverse Female User Groups to Gather Insights: Conduct interviews, surveys, and codesign workshops with women from varied backgrounds, including different ages, cultures, and abilities, to ensure their voices shape the design.

1.12.2.   Build Features That Reflect the Nuanced Needs of Female Users: Develop functionalities that address specific pain points, preferences, and priorities identified through user research, fostering trust and inclusivity in the design.

1.12.3.   Incorporate Empathy-Driven Design Principles: Ensure that the design process actively considers the emotional, psychological, and social contexts of users, creating solutions that resonate on a personal level.

1.12.4.   Use Inclusive Language and Visuals: Avoid stereotypes in text, images, and design elements, ensuring that all users feel represented and respected.

1.12.5.   Address Intersectionality: Recognize that women's experiences vary based on factors like race, socioeconomic status, disability, and sexual orientation, and reflect these complexities in the design.

1.12.6.   Iterate and Validate With User Feedback: Continuously test prototypes with real users to refine the design based on their feedback and ensure it meets their expectations and needs.

1.12.7.   Prioritize Privacy and Security in Sensitive Contexts: Especially for features related to personal or health data, ensure that privacy is safeguarded to build user trust and protect their dignity.

1.12.8.   Provide transparency in design decisions: Communicate the purpose and intent behind features or changes to users, showing respect for their input and fostering a sense of collaboration.

Accessibility, in the context of practicing data feminism, is more than a design principle; it is a commitment to equity and justice in technology. By embracing universal design principles (Centre for Excellence in Universal Design, 2025) and integrating assistive technologies, designers and developers of artificial intelligence-powered data systems can create platforms that empower individuals with diverse abilities. The Centre for Excellence in Universal Design (2025) emphasizes that accessibility must be proactive, ensuring that digital tools are usable by all from the outset, rather than requiring retroactive fixes. This approach not only enhances quality of life but also fosters greater independence, economic inclusion, and social participation. Moreover, accessible design strengthens data integrity by ensuring that a wider, more representative user base can engage with and contribute to technological ecosystems. This reference aims to inspire the creation of reproductive health platforms that are both inclusive and impactful, reflecting the values of accessibility, equity, and user-centered innovation.

CHAPTER 3

User Consent

Ensuring Informed Consent in Equity-Centered Design for Women's Reproductive Health Data

Informed **user consent** is a cornerstone of equity-centred design, particularly in the context of sensitive personal health data related to women's reproductive health. This heuristic category underscores the necessity of clear and explicit consent, ensuring women fully understand the scope of data collection, usage, access, and protection. By integrating thoughtful information design and **empathetic characterization**, designers can empower women to make informed decisions, fostering transparency and respect for user autonomy. This section provides actionable heuristics and real-life applications for integrating informed consent in reproductive health platforms.

UNDERSTANDING INFORMED CONSENT IN EQUITY-CENTERED DESIGN

Informed consent refers to the process of obtaining explicit, voluntary, and fully informed permission from individuals for the collection, processing, and use of their personal data. It is a fundamental principle in equity-centred design, prioritizing the protection of user autonomy, dignity, and privacy.

Reproductive health data are deeply personal and often associated with societal and cultural stigmas, making their protection a critical ethical obligation. Informed consent is especially important to safeguard women's dignity and prevent misuse of their sensitive data.

Equity-centred design demands informed consent that goes beyond perfunctory acceptance:

- Users must have access to complete, comprehensible information about data practices.

- The consent process should empower users to make deliberate, confident decisions.

Key Elements of Informed Consent

Effective informed consent should address the following:

- **Data Collection:** What data are being collected and why

- **Purpose of Use:** How the data will be used, whether for research, analytics, or services

- **Access Control:** Who will have access to the data

- **Protection Measures:** How the data will be secured against breaches or unauthorized access

APP DESIGN AND EMPATHETIC CHARACTERIZATION

Empathetic characterization, in the context of application design, involves more than just acknowledging the unique needs, experiences, and concerns of female users—it demands a deep ethical commitment to representing them accurately, respectfully, and without distortion. When technology mischaracterizes or reduces individuals to data points stripped of context, it risks reinforcing harmful biases, erasing lived experiences, and perpetuating systemic inequalities. True empathetic characterization requires active listening, intersectional awareness, and the intentional design of systems that honor user agency and dignity. In reproductive health applications, this means moving beyond generic, one-size-fits-all solutions and instead embedding representation that is nuanced, just, and reflective of the diverse realities of women's lives:

- Apps should foster an environment of trust and respect through intuitive and empathetic design.

- This includes tone, visuals, and features that resonate with the diverse experiences of women.

Information design simplifies complex data practices into understandable formats:

- Use **plain language** to explain consent terms.

- Incorporate visuals, infographics, and interactive tools to illustrate data processes clearly.

- Present information incrementally to avoid overwhelming users.

In the wake of the *Roe v. Wade* reversal in the United States, period-tracking apps like Flo, introduced in Chapter 2,[1] faced

---

1 You will note the same companies in many of the examples cited.

significant backlash for failing to adequately inform users about how their data might be used. Although Flo marketed itself as a trusted reproductive health tool, it was revealed that the app had shared sensitive data with third parties, including analytics and marketing companies, without clear or explicit user consent.

In January 2021, the Federal Trade Commission (FTC, 2021a, 2021b) reached a settlement with Flo over allegations that the company misled users about the privacy of their health data. Despite explicit assurances that sensitive health information would remain confidential, Flo allegedly shared user data, including pregnancy status, with third-party analytics and marketing firms such as Facebook and Google—without obtaining proper user consent. This practice persisted from 2016 until it was publicly exposed in 2019, prompting widespread criticism and user complaints. As part of the settlement, Flo was required to obtain affirmative user consent before sharing health information, undergo independent privacy assessments, notify affected users of prior unauthorized disclosures, and instruct third parties to delete improperly shared data. The FTC also prohibited the company from misrepresenting its data collection, user control over personal information, and compliance with privacy standards. This case serves as a stark reminder of the critical need for transparency, accountability, and regulatory oversight in digital health platforms that handle intimate and highly sensitive user data.

Sadly, many women, unaware of the potential risks, had entered deeply personal information, only to discover that their data could potentially be used in legal investigations or exploited in ways they had not anticipated. This lack of transparency eroded user trust and exposed the critical need for stronger informed consent mechanisms in apps handling women's reproductive health data. Per earlier mention of this well-documented case, this lawsuit was incited by a U.S. FTC decision where Flo Health, the developer of the app, admitted it had sent Flo users' private information about their periods and pregnancies to data analytics divisions of Google, Facebook, and two other firms— and it did so without informed consent. The company has since updated its policy pages (Flo Health, 2024) to make this information more accessible to its customers.

Nonetheless, the successful implementation of this heuristic category ensures the collection of informed consent from the outset.

When a user signs up for an app like Flo, they should be presented with a clear and concise explanation of how their data will be used, who will have access to them, and for what purposes. The app should use plain language (e.g., "We use your data to provide personalized health recommendations and will never share it without your permission") and offer a detailed, easy-to-access breakdown for those seeking more information.

Learning from their errors, the Flo platform now allows users to opt in or out of specific data-sharing practices, such as contributing anonymized data to research studies, and periodically prompts them to review and update their consent settings. For sensitive areas like reproductive health, the app now provides context-specific consent prompts—for example, explicitly asking for permission before sharing any fertility or pregnancy data with a health care provider (Flo Health, 2024).

Such a model ensures that women are fully informed and in control of their personal information. By prioritizing transparency and user autonomy, the app builds trust while fostering ethical data practices that respect women's rights and privacy in highly sensitive contexts.

In practice, the following guidance should prevail as we think about the agency of women in consenting to the use of their data.

- **Consent Process:** The platform incorporates a step-by-step consent process that details what data are collected, why they are needed, and how they will be used.

- **Interactive Options:** Users can opt in or opt out of specific data-sharing practices, such as research participation or data analytics.

- **Security Assurance:** The platform prominently displays security features and data protection measures to reassure users of their privacy.

Lastly, it cannot be overstated that companies that do not adhere to consent policies should be held accountable through strict penalties, regulatory oversight, and potential legal action to ensure user rights, privacy, and trust are consistently protected. Flo's experience should serve as a testament to this principle.

Proposed Heuristics for Informed Consent in System
Design That Consider Women's Rights and Equity

Informed consent is foundational to ethical system design, ensuring users understand and actively agree to how their data are collected, used, and shared. However, traditional consent models often overlook the unique vulnerabilities and lived experiences of women, reinforcing power imbalances and limiting agency. Designing for equitable informed consent requires addressing gendered risks, promoting transparency, and empowering users to make decisions that reflect their autonomy and rights.

The following heuristics offer a framework for integrating informed consent practices that prioritize women's rights and equity. By embedding clarity, choice, and intersectional sensitivity into system design, these guidelines seek to protect users from exploitation and foster greater trust, safety, and inclusivity in technological environments.

1.13. Consent Forms and Explanations

1.13.1.   Provide Detailed, Jargon-Free Consent Forms: Clearly outline data collection, usage, storage, and sharing practices in a manner that is transparent and easily understandable.

1.13.2.   Use Concise, User-Friendly Language: Ensure the information is accessible to users with varying literacy levels, avoiding technical terms and unnecessary complexity.

1.13.3.   Highlight Key Points Up Front: Summarize critical aspects of consent, such as what data are collected, why they are needed, and how they will be used, before presenting detailed explanations.

1.13.4.   Offer Multilanguage Support: Provide consent forms and explanations in multiple languages to accommodate diverse user populations.

1.13.5.   Incorporate Visual Aids: Use icons, charts, or videos to clarify complex concepts and enhance user comprehension.

1.13.6. Allow Granular Consent Options: Enable users to opt in or out of specific data practices (e.g., data sharing with third parties) rather than using an all-or-nothing approach.

1.13.7. Provide Opportunities for Clarification: Include accessible resources such as FAQs, help centers, or live chat options for users who need more information before consenting.

1.13.8. Ensure Revocability of Consent: Allow users to withdraw consent at any time, with clear instructions on how to do so and what the implications might be.

1.13.9. Conduct Usability Testing on Consent Forms: Test consent forms with diverse user groups to ensure they are truly comprehensible and meet the needs of all users.

1.13.10. Adhere to Legal and Ethical Standards: Align consent processes with regulations such as the General Data Protection Regulation, the Health Insurance Portability and Accountability Act, or other relevant frameworks, ensuring compliance and user protection.

1.14. Opt-In and Opt-Out Mechanisms

1.14.1. Offer clear, intuitive mechanisms that allow users to opt in or opt out of specific data uses, ensuring they understand the implications of their choices.

1.14.2. Provide the ability for users to withdraw consent and delete their data at any time, with straightforward processes and transparent explanations of the outcomes.

1.14.3. Ensure users are notified of any changes to data use policies and given the opportunity to adjust their preferences accordingly.

1.14.4. Design mechanisms that respect user decisions without penalizing them, such as restricting access to unrelated features if they choose to opt out of certain data uses.

    1.14.5.   Incorporate reminders or periodic prompts for users to review and update their data preferences, fostering ongoing consent.

1.15. Transparent Data Usage

    1.15.1.   Be explicit about how data will be used, providing clear explanations of its purposes and benefits to the user.

    1.15.2.   Inform users if their data will be shared with third parties, detailing who the recipients are, why the data are shared, and securing additional consent for such activities.

    1.15.3.   Provide real-time visibility into how user data are being processed or accessed, such as through a dashboard or activity log.

    1.15.4.   Communicate any changes to data usage policies in a timely manner and allow users to review and modify their preferences as needed.

    1.15.5.   Ensure all data usage practices align with legal regulations and ethical standards, prioritizing user trust and autonomy.

1.16. Data Protection and Security

    1.16.1.   Highlight the security measures in place, such as encryption, secure servers, multifactor authentication, and regular system monitoring, to reassure users their data are protected.

    1.16.2.   Provide regular updates about system security, including notifications of improvements or changes, to build user trust and demonstrate a commitment to safeguarding their information.

    1.16.3.   Ensure compliance with international and local data protection regulations, such as the General Data Protection Regulation or the Health Insurance Portability and Accountability Act, to meet legal standards and enhance user confidence.

1.16.4. Enable users to easily access information about security features and protocols, offering transparency about how their data are being safeguarded.

1.16.5. Proactively address potential vulnerabilities through regular security audits, penetration testing, and timely patches or updates to mitigate risks.

1.16.6. Establish a clear, user-friendly protocol for reporting security breaches, including steps users can take to protect themselves and how the organization will respond.

1.17. **Ethical Data Sharing**

1.17.1. Seek additional, specific, and informed consent before sharing user data with external parties, such as researchers, ensuring users fully understand the purpose and scope of the data sharing.

1.17.2. Adhere strictly to ethical guidelines by implementing robust measures like **data anonymization,** pseudonymization, and identity protection to safeguard user privacy.

1.17.3. Limit shared data to only what is necessary for the stated purpose, avoiding unnecessary or excessive data disclosure.

1.17.4. Provide users with clear information about who will access their data, why they are being shared, and how they will be used, fostering transparency and trust.

1.17.5. Establish agreements with external parties to ensure shared data are handled in compliance with legal and ethical standards, preventing misuse or unauthorized access.

1.17.6. Allow users to review and revoke consent for data sharing at any time, empowering them to maintain control over their information.

1.18. User Feedback and Involvement

1.18.1. Seek additional, specific, and informed consent for sharing user data with external parties, such as researchers, ensuring users clearly understand the purpose and implications.

1.18.2. Adhere strictly to ethical standards, including implementing data anonymization and identity protection measures to safeguard user privacy.

1.18.3. Ensure data sharing is purpose-driven, limiting the shared information to only what is essential for the stated objectives, and avoiding unnecessary exposure.

1.18.4. Maintain transparency by providing users with detailed information about who will access their data, why they are being shared, and how they will be used.

1.18.5. Establish legally binding agreements with external parties to ensure they comply with ethical guidelines and legal requirements for data handling and protection.

1.18.6. Enable users to track, review, and revoke their consent for data sharing at any time, giving them full control over their personal information.

Informed consent is not just a procedural formality; it is a non-negotiable pillar of equity-centered design, particularly in the deeply personal and politically fraught domain of women's reproductive health data. Without meaningful consent, digital health platforms risk perpetuating the very structures of control and surveillance that feminist movements have long fought to dismantle. By employing empathetic characterization, clear information design, and user-centric privacy controls, platforms can ensure that women are not merely data subjects but active agents in managing their own health information. These principles go beyond ethical best practices; they represent a fundamental commitment to bodily autonomy, data sovereignty, and resistance against exploitative digital infrastructures. In a landscape where reproductive rights are under siege, equity-centered design must not be an afterthought; it must be the standard. This model of transparency, dignity, and user empowerment is not just an aspiration but a moral and political imperative in the responsible design of reproductive health technologies. As I close this chapter, the urgency of this moment could not be more clear, with the current political climate making these discussions more relevant than ever.

CHAPTER 4

Privacy by Design

Proactive Privacy Integration in
Design: Fostering Algorithmic
Trust for Women's Sensitive
Health Data

In today's digital race, where health data are both a prized asset and, concurrently, a profound vulnerability, **Privacy by Design (PbD)** stands not just as a guiding principle but as a moral obligation in ethical and effective technology development. This categorical heuristic, therefore, demands a proactive, anticipatory approach to privacy—embedding safeguards into the design process from the outset, rather than reacting to violations only after harm has occurred. We saw the consequences of failing to do so in the previous chapter, with the case of Flo, where privacy assurances crumbled under the weight of undisclosed data-sharing practices (see Federal Trade Commission, 2021a, 2021b). For women's sensitive health data, PbD is not merely a best practice; it is an ethical imperative, one that determines whether digital tools will empower or exploit, protect or commodify.

By integrating empathetic, characterization-driven design, this approach ensures that privacy protections are not abstract technical features but lived commitments to honoring the integrity of women and their data. It fosters trust—not just among women but within the broader health care ecosystem—by recognizing that reproductive health data are more than numbers; they are deeply personal narratives of agency, autonomy, and vulnerability. A system built with this understanding does not simply offer confidentiality—it affirms dignity, respect, and the fundamental right to control one's own body in an era where data have become a battleground for power.

UNDERSTANDING PROACTIVE PRIVACY INTEGRATION IN DESIGN

Proactive privacy integration requires embedding privacy considerations into the design process from the outset, rather than retrofitting protections after development. This forward-thinking approach anticipates risks and implements safeguards to prevent potential breaches, ensuring that privacy remains a core design principle rather than an afterthought. This is particularly crucial for women's reproductive health data, which are deeply personal, often stigmatized, and vulnerable to misuse. Given the increasing politicization of reproductive rights, heightened vigilance in handling such data is not just a technical necessity but a moral and ethical obligation. Proactive privacy measures uphold user dignity, foster trust, and shield against unauthorized access, ensuring that women can engage with digital health platforms without fear of surveillance, exploitation, or discrimination.

Characterization and Respecting Data Integrity

Characterization—a design approach grounded in empathy and user-centricity—offers essential insights into the lived experiences of women, ensuring that digital systems reflect their realities rather than reducing them to abstract data points. This methodology of observing ethos (character) goes beyond technical robustness; it demands that data protection measures are intuitively aligned with the nuanced concerns, expectations, and vulnerabilities of women. By integrating characterization-driven design, platforms can not only safeguard privacy but also affirm agency, fostering a digital environment where women feel both seen and secure. Trust, in this context, is not a byproduct of compliance but can be seen as an ethical commitment baked into the design itself. To achieve this, I propose a two-tiered framework for trust: algorithmic trust and data trust—one ensuring that AI-driven decisions are fair, unbiased, and explainable and the other guaranteeing that users' data remain confidential, protected, and under their control.

- **Fostering Algorithmic Trust:** Generally, **algorithmic trust,** as the name suggests, refers to the confidence health care providers place in the accuracy, reliability, and ethical use of data-driven tools. When algorithms are designed with privacy embedded, health care professionals can trust the outcomes derived from sensitive health data.

- **Establishing Data Trust: Data trust,** on the other hand, goes beyond technical security—it is the confidence users have that their personal data are handled with care, respect, and adherence to ethical standards. Trustworthy systems empower women to engage with platforms without fear of exploitation or misuse.

A compelling example of the importance of PbD is seen in mobile health apps tailored to women's reproductive health, such as those for cycle tracking or fertility management. When privacy is overlooked, the consequences can be severe, as illustrated by the case of the Glow fertility app (https://glowing.com/).

Glow became a cautionary tale when researchers discovered critical security flaws that exposed sensitive user data, including menstrual cycles, sexual activity, and pregnancy information, due to inadequate encryption and weak privacy safeguards (Beilinson, 2020). In 2020, the California attorney general reached a settlement with the company, highlighting privacy violations and unauthorized access risks (Nahra et al., 2020). And most recently, a 2024 security breach exposed personal data of approximately 25 million users, further underscoring ongoing concerns about Glow's data security practices (Franceschi-Biccierai, 2024). These incidents emphasize the need for a more robust data protection in health-related applications.

The fallout extended beyond technical failures. The breach eroded user trust, demonstrating how neglecting privacy not only jeopardizes sensitive information but also undermines the credibility of the platform itself. The Glow case underscores the ethical and operational necessity of embedding privacy protections at the earliest stages of system development, ensuring reproductive health platforms prioritize user safety from inception.

A standout example of PbD is the approach taken by the secure health platform Clue (2025; see also Walter & Tsang, 2022), a menstrual tracking app. Clue explicitly commits to never selling user data, incorporates end-to-end encryption for sensitive health information, and ensures that all data processing is fully compliant with international standards like the General Data Protection Regulation (GDPR).[1] The app's privacy policy is transparent, using clear, accessible language to explain how user data are managed.

Clue prioritizes privacy by embedding robust features directly into its design. These include local data storage options, enabling users to store data on their devices rather than external servers, and tools that allow for the permanent deletion of user data at

---

1   The GDPR is a regulation enacted by the European Union to protect the privacy and personal data of individuals within the EU. It sets guidelines for the collection, storage, and processing of personal data, requiring organizations to obtain explicit consent from individuals and ensuring individuals have rights to access, correct, and erase their data.

any time. The company's female co-CEOs remain actively engaged with the evolving landscape of women's data privacy, particularly in the post–*Roe v. Wade* era, demonstrating a commitment to protecting user rights in a rapidly changing environment (Walter & Tsang, 2022).

These proactive measures not only protect sensitive health data but also foster algorithmic trust, ensuring users feel confident that the recommendations and insights they receive are derived ethically and securely. By integrating robust privacy safeguards from the outset, Clue exemplifies how proactive design can protect women's sensitive health data while reinforcing user trust and upholding ethical standards in digital health technology.

To replicate the Clue example, designers of such a platform should do the following:

- Integrate privacy features from the initial development stages.

- Implement secure encryptions for all stored and transmitted data.

- Design user interfaces that prioritize transparency and control over privacy settings.

By taking these steps, an app not only protects sensitive data but also fosters trust among its users and the health care providers relying on its insights.

Proposed Heuristics for Privacy in System Design That Consider Women's Rights and Equity

PbD is a fundamental heuristic for artificial intelligence (AI) and other emerging technologies that require designers to go beyond traditional user experience considerations to understand how data management informs system design. While designers are not expected to implement technical solutions themselves, they play a critical role in advocating for privacy and security by embedding these principles into the user experience. This collaborative approach ensures that security features are seamlessly integrated into the design process and resonate with users, fostering trust and transparency. The following are best-practice heuristics for designers to consider when applying PbD principles:

1.19. Data Encryption and Storage

1.19.1. Advocate for Secure Data Handling: Ensure that data handled by your system are protected using end-to-end encryption, following industry standards like Advanced Encryption Standard (AES) 256-bit for secure transmission. Transport Layer Security (TLS)[2] should also be implemented to further safeguard data during transit. Consider how these security measures can be communicated visually in the user interface to reassure users.

1.19.2. Incorporate Secure Storage Principles in Design: Champion the use of secure storage methods, such as password hashing (e.g., Argon2 or bcrypt)[3] and data masking, to protect sensitive user information. While not implementing these directly, understand their importance and ensure your designs accommodate their integration.

1.19.3. Promote Safe Key Management: Collaborate with technical teams to include encryption key management practices, such as using hardware security modules or trusted key vaults. Advocate for clear design elements that highlight secure handling of sensitive data.

1.19.4. Design for Compliance and Auditing: Advocate for regular reviews of encryption and storage practices to ensure they align with evolving regulations like the GDPR, Artificial Intelligence and Data Act, Health Insurance Portability and

---

2 AES 256-bit is a widely used encryption algorithm considered highly secure for protecting data during transmission. TLS is a cryptographic protocol designed to ensure privacy and data integrity between applications communicating over a network, such as when transmitting sensitive information over the Internet.

3 Argon2 and bcrypt are two examples of advanced hashing algorithms designed to resist brute-force attacks. Data masking refers to the process of obfuscating sensitive information to prevent unauthorized access while allowing necessary data operations. While not implementing these directly, understand their importance and ensure your designs accommodate their integration.

Accountability Act, and California Consumer Privacy Act (CCPA). Consider how your design might support transparency and compliance reporting, such as dashboards or audit logs.

1.19.5. Build Trust Through Communication: Incorporate elements in user-facing designs that explain encryption practices in clear, user-friendly language. For example, use icons, tooltips, or text that emphasize how user data are being securely managed.

1.20. Secure Authentication

1.20.1. Embed Advanced Authentication Options: Design interfaces to support secure authentication techniques such as multifactor authentication or biometric login. Ensure these options are visible and easy to use, while providing clear explanations of their benefits.

1.20.2. Prioritize Accessibility in Authentication Design: Offer alternative authentication methods to meet diverse user needs, such as hardware tokens, backup codes, or passwordless options. Include design elements that guide users in selecting the method most appropriate for them.

1.20.3. Advocate for Secure Credential Practices: Work with technical teams to avoid insecure practices like storing raw passwords or biometric data. Design for hashed password storage and secure credential management, ensuring this is communicated effectively in the user interface, the medium with which the user interacts.

1.20.4. Design for Upgradability: Create authentication flows that can evolve with emerging threats and technology, such as **adaptive authentication** techniques. Ensure designs accommodate seamless updates without disrupting the user experience.

1.20.5. Visualize Security: Use design elements to highlight authentication security measures, such as verification badges or clear progress indicators, to build confidence in the system's safeguards.

1.21. Minimization of Data Collection

1.21.1.   Collect only the data essential to the platform's functionality, ensuring no unnecessary information is gathered.

1.21.2.   Avoid collecting sensitive or personally identifiable information unless it is strictly required for core operations.

1.21.3.   Design systems to support **data minimization** principles by default, such as anonymizing or aggregating data where possible.

1.21.4.   Regularly review data collection practices to identify and eliminate nonessential data collection.

1.21.5.   Clearly communicate to users what data are being collected, why they are needed, and how they will be used to foster transparency and trust.

1.22. Data Anonymization

1.22.1.   Remove or obscure personal identifiers in data to protect user identities, especially when data are used for research or analytics.

1.22.2.   Use techniques such as pseudonymization, data masking, or generalization to minimize the risk of reidentification.

1.22.3.   Regularly test anonymized datasets to ensure that reidentification risks remain low, particularly as data and technology evolve.

1.22.4.   Follow established frameworks and standards for anonymization to ensure compliance with regulations such as the GDPR or CCPA.

1.22.5.   Clearly document anonymization practices to provide transparency and accountability in how user data are handled.

1.23. Regular Security Audits

1.23.1.   Conduct periodic security assessments to identify vulnerabilities in systems, applications, and data-handling processes.

1.23.2. Implement updates and patches promptly to address any security issues uncovered during audits.

1.23.3. Include penetration testing, code reviews, and vulnerability scans as part of the security audit process.

1.23.4. Maintain detailed records of audit findings, actions taken, and improvements implemented to track progress and compliance.

1.23.5. Stay informed about emerging security threats and adjust audit protocols to address new risks proactively.

1.24. **Transparent Privacy Policies**

1.24.1. Provide a clear and concise privacy policy that is easy for users to understand.

1.24.2. Clearly explain what data are collected, specifying the types of information and the reasons for collection.

1.24.3. Describe how collected data will be used, including any potential secondary purposes or data-sharing scenarios.

1.24.4. Identify who will have access to the data, including internal stakeholders and external parties, such as third-party service providers.

1.24.5. Detail the measures in place to protect user data, including encryption, access controls, and compliance with relevant regulations.

1.24.6. Regularly review and update the privacy policy to reflect changes in data practices or regulatory requirements.

1.24.7. Make the privacy policy easily accessible across platforms to promote user trust and transparency.

1.25. **Ethical Considerations**

1.25.1. Take a characterization-driven approach to ensure ethical handling of sensitive health data, with particular attention to the unique privacy needs and concerns of women.

1.25.2. Design systems and policies that respect cultural, societal, and individual sensitivities related to health data.

1.25.3. Incorporate user feedback into the development of privacy features to ensure designs align with user needs, values, and expectations.

1.25.4. Conduct regular evaluations of ethical practices to identify and address potential biases or unintended consequences in data handling and system design.

1.25.5. Foster transparency and accountability by openly communicating how ethical considerations are integrated into privacy and data management practices.

1.26. Trustworthiness Indicators

1.26.1. Transparency: Ensure trust indicators clearly communicate the organization's adherence to privacy and security standards, making it easy for users to understand their significance.

1.26.2. Recognition: Use certifications and seals from widely recognized and respected organizations within the AI and tech ecosystem (e.g., ISO 27001,[4] GDPR compliance, Ai.ethics Recognition Seals[5]).

1.26.3. Contextual Placement: Embed trust indicators in user-facing interfaces where trust may be most needed, such as during data collection, payment processes, or AI decision disclosures.

---

4   ISO is an internationally recognized standard for information security management systems.

5   Initiatives like the Ai.ethics Recognition Seals (https://aiethics.pt/en /recognition-ai-ethics/) are awarded to organizations that demonstrate ethical governance in their AI systems, aiming to instill trust among stakeholders.

1.26.4. Consistency: Apply trust indicators consistently across interfaces to reinforce a cohesive trust experience.

1.26.5. Authenticity: Regularly audit and verify the authenticity of displayed trust indicators to maintain credibility and avoid misleading users.

1.26.6. Future-Focused: Where applicable, explore emerging trust mechanisms such as blockchain-backed certifications or explainability frameworks to remain at the forefront of ethical AI design.

Proactive privacy integration is essential in limiting how data construct and represent women's identities, particularly in sensitive health care contexts. PbD ensures that reproductive health platforms embed safeguards that prevent data from becoming a reductive proxy for the individual, acknowledging that data are not the person. This approach mitigates the risks of misrepresentation, exploitation, or surveillance, reinforcing that digital traces should never compromise a woman's material existence or autonomy. By embedding privacy at the core of design, developers uphold ethical responsibilities, fostering trust and ensuring that women's health data are treated with the nuance, care, and respect warranted.

CHAPTER 5

Human Factors and Ergonomics

Foundational Guidance
for Designing Women's
Health Technology

This heuristic guidance focuses on the pivotal role of **human factors and ergonomics (HF&E)** in designing technology with consideration for women's health. It highlights the necessity of creating user-friendly, comfortable, and secure interfaces and associated experiences that address the cognitive, emotional, and physical needs of female users. Technology, designed with this categorical heuristic guidance in mind, can not only empower women but also build trust and foster a sense of security. The subheuristics within this category, therefore, aim to align design principles with the diverse and unique requirements of women's health, ensuring inclusive solutions that respect privacy and provide significant, meaningful benefits.

This subsequent section provides an in-depth exploration of HF&E concepts, supported by actionable recommendations and a real-life example of their application in mobile health technologies.

THE IMPORTANCE OF HUMAN FACTORS AND ERGONOMICS IN WOMEN'S HEALTH

HF&E is a discipline dedicated to optimizing interactions between humans and systems to improve efficiency, safety, and overall user experience. When applied to women's health technology, HF&E ensures that systems are not only functional but also intuitive, accessible, and responsive to the unique needs of women as a diverse user group. From medical devices to mobile health apps, well-designed technology can enhance health outcomes, promote ease of use, and build trust by addressing the specific physiological, cognitive, and emotional factors that influence women's interactions with technology.

One critical aspect of women's health technology is the handling of highly personal and sensitive data. Women's health applications frequently process information related to menstrual cycles, fertility, pregnancy, menopause, and other reproductive health concerns. Given the deeply private nature of these data, systems must be designed with privacy, security, and user control in mind. For example, a fertility tracking app should provide encrypted data storage, customizable privacy settings, and clear consent mechanisms to ensure users feel secure when inputting and sharing their information. Without proper HF&E considerations, users may feel reluctant to engage with these tools, ultimately reducing the effectiveness of the technology.

Additionally, women have distinct cognitive, physical, and emotional characteristics that must be factored into system design. Physically, women may have different ergonomic requirements due to variations in hand size, grip strength, and biomechanics, which can affect the usability of medical devices, such as blood pressure monitors or wearable fitness trackers. Cognitively, women may engage with technology differently, often preferring intuitive interfaces with clear, concise instructions and visually accessible displays. Emotionally, the design of health technology should be empathetic and supportive, particularly in areas such as maternal health, mental wellness, and chronic illness management, where stress and emotional burden may be high. By integrating HF&E principles into the development of women's health technology, designers can create systems that respect privacy, accommodate diverse physiological and psychological needs, and ultimately improve the overall health care experience for women.

Respecting Privacy and Safety

Designers must acknowledge the heightened concerns women face regarding data security and privacy in today's digital landscape. Following the reversal of *Roe v. Wade* in the United States, women's advocacy groups have raised significant concerns about how their data are being collected, stored, and used. Two years after the U.S. Supreme Court's decision to overturn *Roe v. Wade*, concerns about data privacy, especially regarding reproductive health information, had begun to intensify (Sherfinski, 2024). Advocates fear that personal data, such as online histories and location information, could be utilized to enforce abortion restrictions. In response, states like Washington have enacted legislation like the My Health My Data Act to bolster protections for health data, reflecting the critical importance of safeguarding personal information in today's data-driven tech economy (Sherfinski, 2024). These concerns have prompted responses from several app creators catering to women's health. For instance, the co-CEOs of the menstrual health app Clue, Carrie Walter and Audrey Tsang (2022), addressed these worries directly:

> We have received messages from our American users concerned about how their tracked data could be used by US courts if Roe vs Wade were overturned. We completely understand the anxiety, and, frankly, the fury that this has even surfaced as a potential risk to worry about. Navigating our reproductive health journey is complex enough, we should never have to wonder whether surveillance of our private patient data could be used to prosecute us.
>
> As the female Co-CEOs of Clue, we promise you that we will never turn your private health data over to any authority that could use it against you. Your personally identifiable health data regarding pregnancies, pregnancy loss or abortion, is kept private and safe. We don't sell it, we don't share it for anyone else's use, we won't disclose it. We are governed by the world's strictest privacy laws (the European GDPR [General Data Protection Regulation]), and we invest a lot of time and expense in making sure we comply with them.
>
> Clue is made to empower you on your reproductive health journey. We aren't here to tell anyone what to do, or what to believe—we aren't here to evangelize for any particular life choice. We are here to support people in the full diversity of their own individual menstrual, sexual and reproductive experience, in every way that trustworthy information can.
>
> Your tracked health data should serve you. We commit to ensuring that it isn't misused for anyone else's agenda.

Prioritizing privacy and safety fosters user trust and confidence. For a mobile health app designed to help women track menstrual cycles, manage reproductive health, and access reliable medical information, like Clue, a successful implementation of HF&E would involve designing interfaces that are intuitive and easy to navigate, accommodating diverse literacy levels and cognitive abilities. It would also prioritize user comfort by ensuring accessible features, such as voice-to-text input, color-blind-friendly visuals, and compatibility with assistive devices. Additionally, the system would integrate robust security measures to ensure sensitive health data remain private and safe. However, real-world examples of where these principles were neglected can be seen in the controversy surrounding Flo, which I reference in Chapters 2, 3, and 4 (Federal Trade Commission [FTC], 2021a, 2021b), and Premom, another fertility app (Piepgrass et al., 2023).

In 2019, Flo was found to be sharing sensitive data with third-party advertisers, including details about women and girls' menstrual cycles and pregnancies, without clear consent. A year later, in 2020, the International Digital Accountability Council raised concerns that Premom had shared sensitive user data with third parties, including two China-based companies flagged for questionable privacy practices (Piepgrass et al., 2023). This data sharing occurred without users' knowledge or consent. Consequently, in 2023, investigations by the District of Columbia, Oregon, and Connecticut, in coordination with the FTC (2021a, 2021b), confirmed these unauthorized data-sharing practices by the company (Piepgrass et al., 2023).

In January 2021, the FTC reached a settlement with Flo for sharing users' sensitive health data with third-party firms, including Facebook and Google, despite promising to keep such information private. As part of the settlement, app developer Flo Health agreed to obtain users' consent before sharing their health information and to undergo an independent review of its privacy practices (FTC, 2021a, 2021b).

Similarly, as in the case of Flo Health, Easy Healthcare Corporation, the company behind Premom, agreed to implement significant changes to its privacy and security programs and paid a $100,000 penalty to the states involved (Piepgrass et al., 2023).

These breaches of trust not only exposed deeply personal information but also highlighted a lack of transparency and disregard for female autonomy. Such design failures demonstrate the critical need for HF&E principles to guide the development of health apps. Without prioritizing usability, privacy, and ethical design, even well-intentioned technology can jeopardize female safety and erode trust. Given this wider heuristic guidance, the following considerations apply, contextually:

- **User Interface Design:** Intuitive navigation, accessible menus, and clear instructions

- **Privacy Features:** Secure logins, encrypted data storage, and user-controlled permissions

- **Inclusive Aesthetics:** Colors and visuals that evoke calmness while avoiding stereotypes, bias, and ideology

- **Physical Ergonomics:** Designing apps, associated tools, and tasks to fit women's needs in terms of reducing injury risk and enhancing comfort, safety, and productivity

Proposed Heuristics for HF&E in System Design
That Consider Women's Rights and Equity

HF&E in system design focuses on creating environments and technologies that align with human capabilities, limitations, and needs. However, standard HF&E approaches often reflect male-centric data, overlooking the diverse physiological, cognitive, and social experiences of women. Established human factors and ergonomics research has established this research, which is recalled again, in the context of data feminist studies. Caroline Criado Perez's book *Invisible Women: Data Bias in a World Designed for Men* (2019) highlights how data bias leads to the design of products and environments that do not account for women's needs, resulting in adverse outcomes. Additionally, research by Londa Schiebinger on gendered innovations emphasizes the importance of integrating sex and gender analysis into research to foster innovation and avoid such biases (see "Gendered Innovations," n.d.; HistoryoScience, 2021). Furthermore, the field of feminist human-computer interaction advocates for inclusive design principles that consider diverse user experiences, challenging traditional male-centric perspectives in technology design (Bardzell, 2010). This exclusion can lead to systems that are less effective, safe, and comfortable for women, reinforcing inequities in technological access and usability.

The following heuristics provide a framework for embedding gender equity into HF&E practices. By prioritizing intersectionality, inclusive data collection, and adaptability, these guidelines seek to ensure that system design accommodates the full spectrum of human diversity, fostering safer, more accessible, and empowering experiences for women and marginalized communities.

1.27. **User Interface Design**

Health care app interface should be simple, intuitive, and easy to navigate, designed with diverse preferences and experiences in mind. Recognizing the unique priorities of women as users, key considerations include the following:

1.27.1.   Clear Visual Elements: Use clear and universally understood icons and labels to minimize confusion. Ensure visual elements are intuitive and accessible, considering factors such as color contrast and font size for readability for diverse groups of female users (age, ability, etc.).

1.27.2.   Logical Layouts: Ensure workflows are seamless and align with the user's natural thought processes, minimizing unnecessary steps or redundant actions. Provide visual cues to guide users through complex tasks.

1.27.3.   Consistency: Maintain consistency in design elements (e.g., buttons, fonts, colors) across the interface to reduce the learning curve and foster familiarity.

1.27.4.   Error Prevention and Recovery: Design interfaces to prevent errors where possible and provide clear, actionable feedback when errors occur, including undo options or help links.

1.27.5.   Accessibility: Follow accessibility guidelines (e.g., Web Content Accessibility Guidelines) to ensure the interface is usable for people with disabilities, including keyboard navigation, screen reader compatibility, and alternative text for images.

1.27.6.   Feedback Mechanisms: Incorporate responsive feedback, such as progress indicators, success confirmations, or error messages, to keep users informed about system status.

1.27.7.   User-Centered Design: Conduct usability test-
          ing to ensure the interface meets representative
          female needs and expectations, iterating based
          on real-world feedback.

1.28. Privacy and Security

Given the sensitive nature of women's health data, robust
privacy and security measures are nonnegotiable.
Technology must prioritize the protection of user data
and systems by doing all of the following:

1.28.1.   Employing encryption protocols to secure data
          storage and transmission

1.28.2.   Utilizing industry-standard encryption to pro-
          tect sensitive information from unauthorized
          access or breaches

1.28.3.   Implementing multifactor authentication for
          secure access

1.28.4.   Requiring multiple layers of verification to
          enhance user account security

1.28.5.   Ensuring compliance with data protection reg-
          ulations and standards

1.28.6.   Adhering to frameworks such as the General
          Data Protection Regulation, the Health
          Insurance Portability and Accountability Act,
          or other relevant local and international data
          privacy laws

1.28.7.   Conducting regular security audits and vulner-
          ability assessments

1.28.8.   Proactively identifying and mitigating potential
          threats through routine evaluations and updates

1.28.9.   Providing users with clear and accessible pri-
          vacy controls

1.28.10. Enabling individuals to easily understand and
          manage their data preferences, consent, and
          permissions

1.28.11. Establishing incident response plans

1.28.12. Preparing for potential data breaches or cyber-attacks with predefined protocols for swift and effective action

1.28.13. Minimizing data collection and retention

1.28.14. Adopting a Privacy by Design approach by collecting only essential data and securely disposing of them when no longer needed

1.29. **Inclusivity and Cultural Sensitivity**

Technology should respect cultural diversity and foster inclusivity by doing all of the following:

1.29.1. Providing language options that are neutral and free of bias

1.29.2. Ensuring translations and text outputs are accurate, inclusive, and culturally appropriate

1.29.3. Avoiding culturally insensitive or stereotypical design elements

1.29.4. Designing interfaces and features that consider diverse cultural norms, avoiding assumptions based on stereotypes

1.29.5. Ensuring accessibility for underrepresented and marginalized communities

1.29.6. Incorporating features that cater to individuals with varying levels of access to technology and resources

1.29.7. Supporting diverse representation in content and imagery

1.29.8. Using inclusive visuals, narratives, and characters to reflect a wide range of identities, cultures, and experiences

1.29.9. Engaging culturally diverse female stakeholders in design and testing

1.29.10. Collaborating with individuals from different cultural backgrounds to identify potential biases or insensitivities early in development

1.30. Physical Ergonomics

Not all technology is app-based, and many devices involve physical interactions that should prioritize women's comfort and ease of use. The following principles guide the design of such interactions to ensure inclusivity and usability. In addition to these heuristics, designers must immerse themselves in understanding the unique health and cognitive needs of women in two key areas:

1.30.1. Emotional Contexts: Create tools that accommodate the emotional nuances of health management, such as stress or anxiety related to reproductive health, in the societal context.

1.30.2. Cognitive Load: Simplify tasks to reduce cognitive effort, especially for users managing chronic conditions or complex medical schedules.

By embedding these HF&E principles into the design process, technologists can develop solutions that genuinely respond to the diverse needs of women. A mobile health app built on these heuristics would not only enhance usability and accessibility but also cultivate trust, empowering women to manage their health with confidence and autonomy. Beyond improving individual user experiences, this approach establishes a gold standard for inclusive, ethical, and effective technological design in women's health care—one that prioritizes safety, equity, and long-term well-being for all.

CHAPTER 6

Transparency

The Significance of Transparency in Women's Health Care Data Design

Transparency is a cornerstone of ethical and user-centred design, particularly in applications managing sensitive women's health care data. In this heuristic category, transparency is emphasized as a principle that ensures interfaces are clear, honest, and free from hidden functionalities or deceptive practices. Given the deeply personal nature of women's health care data, transparency is essential to build trust, respect user autonomy, and uphold the highest standards of data privacy and security. By embedding transparency throughout the design process, developers can foster a more ethical and trustworthy digital environment.

UNDERSTANDING TRANSPARENCY IN DESIGN

Transparency in design refers to the unambiguous presentation of information about data collection, processing, and usage. It ensures users have full visibility into how their data are handled and can make informed decisions about their participation. Women's health care data are highly sensitive, often encompassing reproductive health information and personal health behaviors. Transparent design provides users with confidence that their data are managed ethically and securely, reinforcing trust and user engagement.

Key aspects of transparency in women's health care data design include the following:

1. **Clear Data Collection Mechanisms**

    - **Explicit Explanations:** Users must receive detailed and understandable explanations of what data are collected and why.

    - **Transparent Data Processing:** Platforms should provide clear visibility into how data are analyzed, stored, and shared.

    - **No Hidden Functions:** Platforms must avoid deceptive practices or hidden functionalities that could compromise user trust or privacy.

2. **Building Trust and User Autonomy**

   - **Trust Through Transparency:** Transparency builds trust by reassuring users that their data are handled responsibly, with safeguards against misuse or unauthorized access.

   - **Respect for User Autonomy:** Providing clear information empowers users to make informed decisions about their data-sharing preferences, fostering a sense of ownership and control.

3. **Ensuring Privacy and Security**

   - **Secure Data Handling:** Robust security measures must be implemented to protect sensitive data from breaches or unauthorized access.

   - **Communication of Data Usage:** Users should be informed about how their data will be utilized, including whether they will be shared with external entities and for what purposes.

   - **Compliance With Regulations:** Platforms should comply with data protection regulations such as the General Data Protection Regulation, Artificial Intelligence and Data Act, Algorithmic Accountability Act, Health Insurance Portability and Accountability Act, and Personal Information Protection and Electronic Documents Act, ensuring the legal and ethical handling of user information.

A health tracking app tailored to women's health care needs exemplifies the importance of transparency. As discussed in several of the preceding chapters, one high-profile example of insufficient transparency occurred with Flo, one of the most popular period and fertility trackers; another case, referenced in Chapter 5, involves the ovulation and period tracking app Premom. In this case, while the app was marketed as a trusted tool for women's health, in 2019 the app developer Flo Health was found liable by the Federal Trade Commission (2021a, 2021b) for sharing sensitive user data, including menstrual cycle and pregnancy information, with third-party companies like Facebook and Google for marketing purposes. The app's privacy policy did not clearly disclose these practices, leaving users unaware of how their personal data were being utilized. This breach of trust led to a settlement with the Federal Trade Commission

(2021a, 2021b), highlighting the critical need for transparency in health tracking apps. And we have seen a similar outcome in the case of Premom, with both companies paying the price for the lack of transparency provided to their customers as to how their data were being used (Piepgrass et al., 2023).

By contrast, the Natural Cycles app, which markets itself as "an FDA cleared, non-hormonal, non-invasive way to take control of your fertility" (Natural Cycles, n.d.), specializes in fertility tracking and has taken steps to enhance transparency in its data practices (Natural Cycles, 2024). The app's designers prioritize clear communication through the following:

- Providing a plain-language privacy policy that outlines what data are collected, how they are used, and who (if anyone) they are shared with

- Including in-app consent prompts that explain specific data usage, such as contributing anonymized information to research projects, giving users the choice to opt in or out

- Offering **real-time notifications** whenever there are updates to data policies, ensuring users remain informed

- Allowing users to easily download and review their data or delete them entirely, reinforcing trust in the platform's commitment to user control

This transparent approach ensures that users understand how their data are handled, fostering confidence in the app while setting a standard for ethical practices in women's health technology. Apps that prioritize transparency not only protect their users but also strengthen their reputation in an increasingly privacy-conscious marketplace.

In this scenario, the app's designers prioritize clear communication about data collection and processing practices. Key elements might include the following:

- **Accessible Privacy Policy:** Explaining how user data are collected, stored, and shared

- **Real-Time Notifications:** Alerting users when their data are accessed or shared

- **User-Controlled Access:** Enabling users to view, edit, or delete their data at any time

By implementing transparency-focused features, an app builds trust while respecting the sensitive nature of its users' health care data.

Proposed Heuristics for Transparent System
Design That Consider Women's Rights and Equity

Transparency in system design is essential for fostering trust, accountability, and user empowerment. However, many systems obscure decision-making processes, data usage, and algorithmic outcomes—reinforcing inequities that disproportionately affect women and marginalized communities. A lack of transparency can perpetuate biases, limit recourse, and erode agency, ultimately exacerbating systemic inequalities.

The following heuristics provide a framework for integrating transparency into system design with a focus on women's rights and equity. By emphasizing clarity, accountability, and inclusive communication, these guidelines aim to create systems that are not only understandable but also responsive to the diverse needs and experiences of women, promoting fairness and ethical engagement across all user interactions.

1.31. **Privacy Policy**

    1.31.1.   Provide a prominently displayed and easily **accessible privacy policy** that outlines the following:

- The types of data collected

- The purposes of data usage

- Any third parties involved in data sharing

1.32. **Consent Screens**

    1.32.1.   Use clear, concise **consent screens** when users sign up or when new features involve data sharing. These screens should detail specific data collection and usage practices.

1.33. **User Notifications**

    1.33.1.   Notify users in real time whenever their data are accessed, shared, or used in significant ways. Notifications should be simple and actionable.

1.34. Data Access Control

1.34.1. Allow users to manage their data, including viewing what has been collected and editing or deleting these data as needed.

1.35. Plain Language

1.35.1. Avoid technical jargon in privacy policies, notifications, and interfaces. Use straightforward language to ensure comprehension.

1.36. Data Security Measures

1.36.1. Clearly communicate the security measures in place, such as encryption and secure storage protocols, to reassure users about data protection.

1.37. Accessibility of Information

1.37.1. Ensure all transparency efforts are accessible to users with varying abilities, offering formats for those with visual or cognitive impairments.

1.38. Ethical Considerations

1.38.1. Provide comprehensive and honest information, avoiding omissions or vague descriptions that could mislead users.

Transparency in women's health care data design is more than a functional necessity—it is a fundamental ethical imperative. In an era where digital health technologies increasingly mediate personal well-being, ensuring that users have full visibility and control over their data is essential to fostering trust, respecting autonomy, and protecting sensitive information from misuse. Women face additional considerations, as their health data often include deeply personal aspects such as reproductive health, fertility tracking, and hormonal patterns—information that, if misused or inadequately protected, can have serious personal, legal, and societal implications. Thoughtful, transparent design practices not only empower women to make informed decisions about their health but also reinforce ethical data stewardship, aligning with regulatory safeguards and human-centered principles. Beyond mere compliance, transparency serves as the cornerstone of an equitable and trustworthy technological ecosystem—one that upholds the dignity, privacy, and rights of women while dismantling systemic biases in health care technology.

Conclusion

As we traverse the frontiers of artificial intelligence (AI) and data-driven innovation, compounded with Big Tech and state interests, the ethical and equitable design of emerging technologies must be recognized not as a technical aspiration but as a moral imperative. This supplement, rooted in critical discourse and theory, activism, and AI design considerations for women's rights, serves as both a reflection and a call to action. It also highlights the often-invisible intersections between technology, gender, and the enduring legacies of systemic inequity, inclusive of race. The frameworks and heuristics proposed throughout this work do more than mitigate harm; they lay the groundwork for a transformative recalibration of the technological landscape—one that calls for both accountability and a means for auditability. Further, it challenges the status quo of surveillance, exclusion, human datafication, and commodification. Informed consent, accessibility, and user agency, for example, are not mere features of ethical design but the very foundations upon which trust, dignity, and justice can be restored and preserved.

Technology now advances at a pace that far exceeds the capacity of policymakers, regulatory bodies, civil society, and other key stakeholders to meaningfully engage with its implications. This widening gap between innovation and oversight underscores the urgency of fostering continuous dialogue that bridges theory with practice. The design of technology cannot be disentangled from the conversations shaping its development, nor can theory exist in isolation from the cultural and political zeitgeist that informs it. Our task, therefore, is not only to build ethical systems but to cultivate the intellectual and communal spaces where these systems are conceived and critiqued.

It is vital to recognize that technology, while appearing to be neutral, is deeply ideologically imbued with the values, biases, and power dynamics of those who create it. As designers, technologists, and policymakers, we are not only architects of digital experiences but custodians of societal progress. The decisions we make today—how we characterize, represent, and safeguard women's data—will echo far into the future, shaping the contours of rights, autonomy, and identity in the digital age.

This supplement underscores that addressing inequity in technology design is not an isolated act but part of a broader movement to decolonize data, democratize access, and humanize the digital spaces we increasingly inhabit. The examples shared throughout these pages illuminate not only the risks of unchecked technological advancement but the boundless potential for AI and data systems to become instruments of liberation, rather than tools of oppression. True innovation lies not in the efficiency of algorithms but in the courage to ask uncomfortable questions:

Whose interests are being served?

Whose stories are being erased?

Whose futures are being shaped without their consent?

By embedding principles of data feminism and human-centered and intersectional design into the core of our practices, we collectively author a new narrative—one where technology exists not to extract, surveil, or control but to amplify, uplift, and restore.

The horizon of possibility remains open, and with it, the responsibility to design futures that reflect our highest aspirations. Let this work serve as a blueprint for those ready to rise to that challenge, ensuring that as technology evolves, so, too, does our collective commitment to justice, equity, and the enduring rights of women and all marginalized communities.

Afterword

DESIGNING FOR EQUITY IN THE AGE OF ARTIFICIAL INTELLIGENCE

As technologists, designers, and other stakeholders, you hold immense power to shape how female data are represented, interpreted, and embedded in design driven by artificial intelligence (AI). The datafication of women's bodies and experiences transcends technical boundaries—it stands as a profound ethical responsibility. Your role in this evolving landscape is not passive; it demands critical thinking, vigilance, and empathy.

By questioning assumptions, identifying biases in data collection, and fostering transparency, you can actively contribute to an AI ecosystem that respects and amplifies the diverse realities of women. This is not just a design challenge—it is a moral imperative that will influence how AI safeguards women's health, autonomy, and rights in the years ahead.

Let this guide mark the beginning of your journey in navigating the intricate intersections of ethics, technology, and gender. The choices you make today will resonate far beyond individual projects, shaping the societal impact of AI on future generations.

Here's how you can take action:

1. **Challenge Assumptions:** Examine how female data are gathered, interpreted, and applied. Acknowledge and interrogate the historical and systemic biases embedded in these processes.

2. **Advocate for Inclusive Standards:** Push for datasets and algorithms that reflect the full spectrum of women's identities, backgrounds, and lived experiences. Representation matters at every stage.

3. **Collaborate Across Disciplines:** Engage with sociologists, gender studies scholars, and community advocates to design systems that prioritize fairness, equity, and intersectionality.

4. **Ensure Radical Transparency:** Demystify data practices for users, granting them the agency to understand and control how their information is utilized.

5. **Commit to Continuous Reflection:** Technology evolves rapidly; so must your awareness. Regularly reassess how your work influences women's representation and well-being in the digital world.

The future of AI-powered design will be shaped by individuals who choose to lead with intention and integrity. Addressing the datafication of female bodies requires courage, care, and a refusal to accept the status quo. By embedding ethical frameworks into your design process, you help build technology that empowers and uplifts women across all communities.

THIS WORK BEGINS NOW

Advocate fiercely. Lead decisively. Design with purpose.

As this exploration of design heuristics concludes, take a moment to reflect on the broader impact of your craft. Design is never neutral. Every interface, algorithm, and system you shape carries the weight of the values you infuse into it. This reality is as empowering as it is sobering.

The preceding chapters unpacked heuristics that encourage transparency, inclusivity, and empathy—tools to confront and dismantle the silent biases often embedded in technology. From interface aesthetics to the structure of data, these heuristics are more than best practices; they represent a steadfast commitment to equity. They are blueprints for challenging systems that marginalize women and underrepresented voices.

As designers and technologists, you are not just building products—you are crafting the digital scaffolding of society. Let the values you choose to embed in that scaffolding reflect a future where technology serves, protects, and empowers everyone equally.

HEURISTICS BEYOND THE SURFACE—
PRACTICING WITH CRITICALITY

To treat heuristics merely as checklists or best practices risks reducing their transformative potential. The heuristics outlined in this work are not static prescriptions but rather living tools that gain meaning only when applied within the context of lived experience, social critique, and ongoing dialogue. Their value lies in their ability to intersect with the pressing conversations of our time—conversations about reproductive rights, pay equity, algorithmic bias, and the representation of marginalized voices in public and digital spaces.

Design heuristics without critical reflection risk reinforcing the very systems they aim to dismantle. For this reason, the heuristics presented here must be practiced with a deep awareness of the power dynamics that shape design environments. They require practitioners to interrogate not only what is being designed but for whom and by whom. This text resists the notion that heuristics alone are sufficient. Instead, it advocates for an ongoing engagement with intersectional thinking—an approach that recognizes the overlapping and compounding nature of gender, race, class, disability, and identity within technological systems.

INTERSECTIONALITY AS THE
LENS OF OUR TIME

Intersectionality offers us more than a theoretical framework; it serves as a necessary compass for design justice. It forces us to confront the reality that the experiences of women are not monolithic—a Black woman navigating health care systems faces different biases than a White woman using the same tools. The same can be said for individuals who are queer, disabled, immigrant, or nonbinary. Designing for women without understanding these intersecting identities risks reinforcing exclusion under the guise of inclusion.

Through the lens of intersectionality, the heuristics in this text are instruments not only of representation but also of resistance. They resist the tendency to flatten diverse experiences into one-dimensional user profiles. They insist that technology reflect the complexity of real life and respond to it with nuance and care.

A LIVING DIALOGUE IN
CRITICAL DESIGN

This is not just a book about heuristics—it is a contribution to the critical design discourse of our time. It belongs alongside movements advocating for design justice, antiracist technologies, and feminist data practices. By grounding design heuristics in the larger dialogue of equity and representation, this work emphasizes that design can no longer afford to be ahistorical or apolitical.

Design happens within the fabric of society, shaped by the forces of the day. The push for ethical AI, the debate around privacy rights, and the demand for more equitable digital landscapes are inseparable from the design processes we engage in. The heuristics presented here, therefore, must evolve as society evolves, reflecting new insights and adapting to emerging struggles.

CARRYING THE WORK FORWARD

This text is not a conclusion—it is an invitation to critique, to iterate, and to extend these heuristics into new domains. The fight for women's rights within design is ongoing, and the heuristics outlined here represent just one facet of a larger movement toward justice.

As designers, technologists, and advocates, we must continue to ask difficult questions: Who is left out? Whose stories are not being told? How can design serve as a force for liberation rather than limitation?

In answering these questions, we take small but vital steps toward systems that honor the full complexity of human experience. And in doing so, we craft not only better designs but a more just and inclusive world.

Thank you for being part of this critical and urgent work.

Discussion Questions

1. How do patriarchal systems influence the design of artificial intelligence (AI) technologies, and what are the implications for women's rights and reproductive health care?

2. What lessons can be drawn from the AI project in Salta, Argentina (discussed in the introduction), regarding unintended consequences in algorithmic design?

3. How can feminist theory and critical data studies contribute to more equitable technological design?

4. In what ways might AI hinder or support women's reproductive rights, especially in the context of shifting legal frameworks?

5. How does algorithmic oppression manifest in emerging technologies, and what can be done to address it in the design process?

6. How can equity-centered design heuristics be applied to mitigate systemic bias in AI systems?

7. What are the limitations of global policies like the General Data Protection Regulation and the Artificial Intelligence and Data Act in addressing gender bias in AI?

8. How does each of the six heuristic categories proposed in the introduction address systemic bias?

9. How can AI be developed as a tool for empowerment and justice, particularly for marginalized communities?

10. What role do practitioners and policymakers play in fostering inclusive design practices that prioritize the rights of women and marginalized groups?

11. What is algorithmic ethopoeia, and how does it influence the portrayal and characterization of individuals in data-driven systems?

12. What are ethotic heuristics, and how can they be applied to build AI systems that reflect ethical considerations and social values?

13. How does datafication shape perceptions of identity and agency, particularly for women and marginalized communities?

14. In what ways does AI contribute to the human characterization process, and what risks does this pose in terms of reinforcing stereotypes or biases?

15. How can designers ensure that data-driven characterization processes reflect diverse, authentic representations of human experiences?

Resources

As the technology and artificial intelligence (AI) landscape continues to evolve, driven by the rapid iteration inherent to technological development, designers and technologists hold a pivotal role in shaping systems that impact lives globally. This influence carries a profound ethical responsibility to embed equity, inclusivity, and ethical considerations into every stage of the design and development process. However, addressing the multifaceted challenges of creating equitable systems is not without its difficulties, particularly in environments where potentially unethical practices may arise.

These resources aim to address common questions that designers and technologists may encounter regarding their ethical responsibilities, the principles they should uphold, and practical actions they can take when confronting unethical situations. Beyond offering practical guidance, these resources also provide tools and references to empower professionals to advocate for ethical practices and to foster a culture of accountability within their organizations to support the development of systems that align with broader societal values and promote sustainable, ethical innovation.

1. **What is my responsibility as a technologist or designer in creating equitable systems?**

   - Your primary responsibility is to ensure that the systems you design are inclusive, fair, and accessible to diverse user groups. This involves conducting user research to understand the needs of marginalized communities; avoiding biases in algorithms, interfaces, and data collection practices; and advocating for human-centered design that prioritizes the well-being and autonomy of all users.

Resources

Constanza-Chock, S. (2020). *Design justice: Community-led practices to build the worlds we need.* MIT Press.

Observatory of Public Sector Innovation. (n.d.). *Ethical OS toolkit.* Organisation for Economic Co-operation and Development. https://oecd-opsi.org/toolkits/ethical-os-toolkit/

2. How can I identify inequities or biases in the systems I design?

To identify inequities or biases, consider auditing datasets for representation and fairness, examining system output patterns of discrimination, and consulting with diverse user groups to validate assumptions.

Resources

Business Software Alliance. (2021). *Confronting bias: BSA's framework to build trust in AI.* https://www.bsa.org/reports/confronting-bias-bsas-framework-to-build-trust-in-ai

O'Neil, C. (2016). *Weapons of math destruction: How big data increases inequality and threatens democracy.* Crown.

3. What steps can I take to ensure inclusivity in system design?

- Use inclusive design practices, such as considering accessibility needs and cultural differences; incorporate feedback from underrepresented groups during all phases of development; and avoid language, imagery, or features that could alienate or harm certain demographics.

Resources

Cunningham, K. (2012). *The accessibility handbook: Making 508-compliant websites.* O'Reilly Media.

Microsoft. (n.d.). *Microsoft inclusive design.* https://www.microsoft.com/design/inclusive/

4. What should I do if I notice potential harm in the technology I'm working on?

- Document your concerns clearly and objectively, including evidence of potential harm; raise the issue with your team, your manager, or an ethics committee, if available; and propose alternatives or modifications to address the issue.

Resources

Center for Humane Technology. (2025). *Articulating challenges. Identifying interventions. Empowering humanity.* https://www .humanetech.com
Markkula Center for Applied Ethics. (2025). Santa Clara University. https://www.scu.edu/ethics

5. **How can I push back if I believe I'm being asked to design an unethical system?**

   • Voice your concerns professionally, referencing ethical guidelines and industry standards; seek allies within your organization who share your values; and, if necessary, escalate the issue to senior management or an external authority.

Resources

Institute of Electrical and Electronics Engineers. (2020). *IEEE code of ethics.* https://www.ieee.org/about/corporate/governance/p7-8 .html
Occupational Safety and Health Administration. (n.d.). *Whistleblower protection program.* U.S. Department of Labor. https://www .whistleblowers.gov

6. **What ethical principles should guide my work in tech design?**

   • **Equity:** Ensure fairness and inclusion for all users.

   • **Transparency:** Be open about how systems work and how data are used.

   • **Accountability:** Take responsibility for the impact of your designs.

   • **Nonmaleficence:** Avoid causing harm to users or communities.

Resources

Association for Computing Machinery. (2018). *ACM code of ethics and professional conduct.* https://www.acm.org/code-of-ethics
Daugherty, P., & Wilson, H. J. (2018). *Human + machine: Reimagining work in the age of AI.* Harvard Business Review Press.

7. **How do I address resistance to ethical practices in my organization?**

   - Use data and case studies to demonstrate the business and social value of ethical practices, build a coalition of like-minded colleagues to advocate for change collectively, and suggest practical steps, such as introducing bias audits or ethics training.

Resources

Hicks, D. (2018). *Leading with dignity: How to create a culture that brings out the best in people.* Yale University Press.

Incorporating Ethics. (2025). *Ethical leadership toolkit.* https://incorpo ratingethics.ca/ethical-leadership-toolkit/

8. **How can I educate myself further on ethical design and equitable systems?**

   - Attend workshops, webinars, and conferences on ethical tech design; enroll in relevant online courses; and engage with literature and case studies on ethics in technology.

Resources

Berkman Klein Center for Internet & Society. (n.d.). *The latest.* Harvard University. https://cyber.harvard.edu

Ng, A. (n.d.). *AI for everyone* [Online course]. Coursera. https://www .coursera.org/learn/ai-for-everyone

9. **What are some warning signs of unethical design practices in my workplace?**

   - Warning signs include lack of transparency about system functions or data use, exclusion of marginalized voices in decision-making processes, and pressure to prioritize profit over user welfare or societal impact.

Resources

Algorithmic Justice League. (2024). *Take action.* https://www.ajl.org

Benjamin, R. (2019). *Race after technology: Abolitionist tools for the new Jim Code.* Polity Press.

10. **What should I do if I feel my concerns are being ignored or dismissed?**

- Seek external advice or resources, such as ethics organizations or legal counsel, document all communications and actions related to your concerns, and consider reaching out to professional networks or whistleblower support groups.

Resources

National Whistleblower Center. (n.d.). *Whistleblower rights under attack: We need your help now.* https://www.whistleblowers.org

Tech Workers Coalition. (n.d.). *Worker power in the tech industry.* https://www.techworkerscoalition.org

Glossary

Accessibility: The practice of designing products, services, and environments to be usable by people of all abilities, including those with disabilities; ensures that everyone, regardless of physical, sensory, or cognitive limitations, can access and interact with digital and physical spaces effectively.

Accessible privacy policy: A clearly outlined, easy-to-find document explaining how user data are collected, stored, and shared, ensuring users fully understand data practices.

Adaptive authentication: A security mechanism that evolves based on emerging threats by adjusting the authentication requirements dynamically, improving user security and experience.

Algorithmic accountability: The principle that designers and developers of artificial intelligence systems should be responsible for the outcomes of their algorithms, ensuring transparency, fairness, and bias mitigation.

Algorithmic bias: Systemic errors in artificial intelligence models that reflect and perpetuate societal prejudices, leading to unequal outcomes, particularly for women and marginalized communities.

Algorithmic ethopoeia: The mathematized representation of individuals through artificial intelligence, often commodified for commercial, political, or law enforcement purposes, reflecting the construction of characterizations based on fragmented data points, impacting user privacy and autonomy.

Algorithmic trust: The confidence that users place in artificial intelligence systems to handle data ethically and securely, built through transparency, accuracy, and consistent communication about data practices.

Artificial intelligence (AI): Refers to computing systems designed to simulate human intelligence by processing data to enable decision-making and generating predictions. (Note AI is not inherently objective; its outputs are shaped by the data it is trained on, the algorithms that structure its learning, and the biases—both explicit and implicit—embedded in its design. As a result, AI has the potential to either mitigate or amplify social inequities, depending on how it is developed, deployed, and governed.)

Capitalistic realism: Refers to a cultural and ideological condition in which the capitalist system is seen as the only viable framework for organizing society; this perspective normalizes market-driven values and practices, making alternative socioeconomic models appear impractical or even unthinkable and underscoring how deeply entrenched capitalist logic shapes our politics, media, and daily lives, effectively stifling critical debates about systemic change.

Capitalistic reductionism: Refers to an analytical approach that simplifies complex human experiences, social structures, and natural phenomena by reducing them solely to their economic value; it involves viewing and evaluating every aspect of life through the lens of market efficiency, profit, and commodification, often overlooking deeper cultural, ethical, and qualitative dimensions.

Characterization: The process by which artificial intelligence systems construct representations of individuals based on data points, often without contextual understanding, leading to biases and systemic inequalities.

Consent screens: Interfaces designed to clearly communicate data collection and usage practices to users, ensuring transparency and informed decision-making.

Data anonymization: Techniques used to protect user identities by removing or masking personal identifiers in datasets.

Data feminism: A framework that critiques the inequities in data collection, analysis, and application, advocating for inclusive and equitable artificial intelligence systems that reflect the lived experiences of marginalized communities.

Data minimization: The practice of collecting only the essential data necessary for a platform's functionality, reducing the risk of breaches and misuse of sensitive information.

Data trust: A legal and governance framework in which data are held and managed by an independent trustee on behalf of a group of stakeholders, ensuring that data are collected, stored, and shared in a way that aligns with ethical principles, privacy regulations, and the interests of those contributing the data; often used to promote transparency, accountability, and equitable access to data, particularly in areas like health care, smart cities, and development of artificial intelligence.

Datafication: The transformation of various aspects of human life into quantifiable data, often leading to oversimplified representations of identity and experience, particularly in marginalized communities.

Empathetic characterization: Designing artificial intelligence systems and digital interfaces that reflect and respect the lived experiences and diverse identities of users, with sensitivity to marginalized communities.

Empathetic design: A user-centered design process that prioritizes understanding the emotions, needs, and experiences of users, fostering trust and inclusion.

Ethotic heuristics: A framework at the intersection of rhetoric and design that advocates for ethical principles aimed at counteracting the misrepresentation and dehumanization of individuals in technological systems; rooted in intersectional and feminist perspectives, ethotic heuristics prioritize dignity, agency, and equity.

Human-centered design: A design philosophy that prioritizes the needs, values, and experiences of users throughout the development process of artificial intelligence systems and digital platforms.

Human factors and ergonomics (HF&E): A multidisciplinary field focused on optimizing the interaction between humans and systems, ensuring safety, efficiency, and well-being; applies principles from psychology, engineering, and design to create environments, tools, and processes that enhance human performance while minimizing errors and discomfort.

Human oversight: The practice of integrating human judgment into the monitoring, evaluation, and management of automated systems and decision-making processes, ensuring that while technology operates with efficiency and scale, it remains aligned with ethical standards, contextual understanding, and human values; this critical process allows individuals to intervene, correct, or adjust actions as needed, safeguarding against errors, biases, or unintended consequences that can arise from relying solely on automated decision-making.

Informed consent: A critical element in artificial intelligence design that ensures users fully understand and agree to the collection, use, and sharing of their data, with the ability to revoke consent at any time.

Intersectional data practices: Data collection and analysis methods that account for overlapping social identities (e.g., gender, race, class), ensuring that artificial intelligence systems address diverse perspectives and experiences.

Plain language: A design principle that ensures privacy policies, consent forms, and user communications avoid technical jargon, using clear and simple language for accessibility.

Privacy by Design (PbD): A proactive approach to embedding privacy into the design and architecture of systems from the outset, rather than as an afterthought, to protect sensitive health data and foster trust.

Proactive privacy integration: The incorporation of privacy measures at the initial stages of system design to ensure continuous protection of sensitive data throughout the product life cycle.

Real-time notifications: Alerts provided to users when their data are accessed, shared, or modified, fostering trust and transparency in digital platforms, enabling user controls.

Systems design: The process of planning and organizing all the parts of a system so they work together seamlessly, which means breaking down a complex challenge into manageable pieces—whether it is a piece of technology, a business process, or any organized structure—and ensuring each part fits and functions effectively within the whole; this approach not only makes systems more efficient but also ensures they meet real-world needs in a practical and meaningful way.

Transparency: A design principle that emphasizes clear communication about how artificial intelligence systems collect, process, and utilize data, fostering trust and enabling informed user decisions.

User-centered privacy design: A design approach that places user control and data protection at the core of technology development, aligning with the principles of Privacy by Design.

User consent: The explicit and informed agreement given by users to collect, process, or share their data; a fundamental principle of privacy and data ethics, requiring clear communication of what data are collected, how they will be used, and the ability for users to revoke or modify their consent at any time.

User controls and affordances: Design principles that focus on providing users with mechanisms to manage and control their data, enhancing autonomy, privacy, and ethical engagement with technology.

References

AP News. (2023, November 13). *Men earn more than women in egalitarian Norway, report finds. But it's on par with Europe.* https://apnews.com/article/cee468c29b539c8c871809cb726b65bd

Asadollahi, A., Ebrahimzadeh Zagami, S., Eslami, S., & Latifnejad Roudsari, R. (2025). Evaluating the quality, content accuracy, and user suitability of mHealth prenatal care apps for expectant mothers: Critical assessment study. *Asian/Pacific Island Nursing Journal*, *13*(9), Article e66852. https://doi.org/10.2196/66852

Atwood, M. (1985). *The handmaid's tale.* McClelland & Stewart.

Azevedo, B., Juarez, T., & Taylor, A. (2023, March 8). *Blog: Women's health movement: A brief history.* Weitzman Institute. https://www.weitzmaninstitute.org/blog-womens-health-movement-a-brief-history/

BabyCenter. (2021). *July 2021 birth club* [Forum discussion]. https://community.babycenter.com/groups/a6772790/july_2021_birth_club

Baderoon, G. (2014). *Regarding Muslims: From slavery to post-apartheid.* Wits University Press.

Bardzell, S. (2010, April 10). Feminist HCI: Taking stock and outlining an agenda for design. In *Proceedings of the SIGCHI conference on human factors in computing systems* (pp. 1301–1310). ACM Digital Library. https://dl.acm.org/doi/10.1145/1753326.1753521

Barocas, S., Hardt, M., & Narayanan, A. (2020). *Fairness and machine learning: Limitations and opportunities.* MIT Press.

Beilinson, J. (2020, September 17). Glow pregnancy app exposed women to privacy threats, Consumer Reports finds. *Consumer Reports.* https://www.consumerreports.org/electronics-computers/mobile-security-software/glow-pregnancy-app-exposed-women-to-privacy-threats-a1100919965

Benjamin, R. (2019). *Race after technology: Abolitionist tools for the new Jim Code.* Polity Press.

Berry, D. (2017). *The price for their pound of flesh: The value of the enslaved from womb to grave in the building of a nation.* Beacon Press.

BioMed Central. (2025). Aims and scope. *BMC Pregnancy and Childbirth*. https://bmcpregnancychildbirth.biomedcentral.com

Boston Women's Health Book Collective. (1973). *Our bodies, ourselves: A book by and for women*. Simon & Schuster.

Branstetter, Z. (2024, September 23). Did a Georgia hospital break federal law when it failed to save Amber Thurman? A Senate committee chair wants answers. *ProPublica*. https://www.propublica .org/article/amber-thurman-georgia-abortion-wyden-emtala

Browne, S. (2015). *Dark matters: On the surveillance of Blackness*. Duke University Press.

Buolamwini, J., & Gebru, T. (2018). Gender shades: Intersectional accuracy disparities in commercial gender classification. *Proceedings of Machine Learning Research*, *81*, 1–15.

Calacci, D. (2022, February 16). The case of the creepy algorithm that "predicted" teen pregnancy. *Wired*. https://www.wired.com/story /argentina-algorithms-pregnancy-prediction/

Cao, J., Laabadli, H., Mathis, C. H., Stern, R. D., & Emami-Naeini, P. (2024). "I deleted it after the overturn of Roe v. Wade": Understanding women's privacy concerns toward period-tracking apps in the post Roe v. Wade era. In *CHI '24: Proceedings of the CHI conference on human factors in computing systems* (Article No. 813, pp. 1–22). Association for Computing Machinery. https://doi.org/10.1145 /3613904.3642042

Cavoukian, A. (n.d.). *Privacy by Design: The 7 foundational principles— Implementation and mapping of fair information practices*. https:// privacy.ucsc.edu/resources/privacy-by-design---foundational -principles.pdf

Centre for Excellence in Universal Design. (2025). *About universal design*. https://universaldesign.ie/about-universal-design

Chen, L., Smith, R., & Taylor, M. (2019). Analyzing the credibility of health information in pregnancy applications: A systematic review. *Journal of Maternal Health Technology*, *12*(3), 45–59. https://doi .org/10.1016/j.mht.2019.03.004

Clue. (2025, January 19). *Clue privacy policy*. https://helloclue.com /privacy

Congressional Record—House. (2017, February 6). Hippocratic Oath. In *Proceedings and debates of the 115th Congress, first session* (Vol. 163, No. 22, p. H998). https://www.congress.gov/115/crec/2017 /02/06/CREC-2017-02-06-pt1-PgH998.pdf

Couldry, N., & Mejias, U. A. (2019). *The costs of connection: How data is colonizing human life and appropriating it for capitalism*. Stanford University Press.

Cox, D. (2022, August 26). How overturning *Roe v Wade* has eroded privacy of personal data. *BMJ*, *378*, Article o2075. https://www.bmj.com/content/378/bmj.o2075

Criado Perez, C. (2019). *Invisible women: Data bias in a world designed for men*. Abrams Press.

D'Ignazio, C., & Klein, L. F. (2020). *Data feminism*. MIT Press.

Declercq, E., Barnard-Mayers, R., Zephyrin, L. C., & Johnson, K. (2022, December 14). *The U.S. maternal health divide: The limited maternal health services and worse outcomes of states proposing new abortion restrictions*. The Commonwealth Fund. https://www.commonwealthfund.org/publications/issue-briefs/2022/dec/us-maternal-health-divide-limited-services-worse-outcomes

Dobbs v. Jackson Women's Health Organization, 597 U.S. 215 (2022). https://www.oyez.org/cases/2021/19-1392

Eubanks, V. (2018). *Automating inequality: How high-tech tools profile, police, and punish the poor*. St. Martin's Press.

Federal Trade Commission. (2021a, January 13). *Developer of popular women's fertility-tracking app settles FTC allegations that it misled consumers about the disclosure of their health data*. https://www.ftc.gov/news-events/news/press-releases/2021/01/developer-popular-womens-fertility-tracking-app-settles-ftc-allegations-it-misled-consumers-about

Federal Trade Commission. (2021b, June 22). *FTC finalizes order with Flo Health, a fertility-tracking app that shared sensitive health data with Facebook, Google, and others*. https://www.ftc.gov/news-events/news/press-releases/2021/06/ftc-finalizes-order-flo-health-fertility-tracking-app-shared-sensitive-health-data-facebook-google

Fleck, A. (2024, November 13). *What's the state of the gender pay gap?* Statista. https://www.statista.com/chart/13182/where-the-gender-pay-gap-is-widest/

Flo Health. (2024, September 6). *Privacy policy*. https://flo.health/privacy-policy

Floridi, L., Cath, C., & Taddeo, M. (2019, October). Digital ethics: Its nature and scope. In C. Öhman & D. Watson (Eds.), *The 2018 yearbook of the digital ethics lab* (pp. 9–17). Springer. https://doi.org/10.1007/978-3-030-17152-0_2

Franceschi-Biccierai, L. (2024, February 13). *Fertility tracker Glow fixes bugs that exposed users' personal data*. TechCrunch. https://techcrunch.com/2024/02/13/fertility-tracker-glow-fixes-bugs-that-exposed-users-personal-data

Frid, G., Bogaert, K., & Chen, K. T. (2021). Mobile health apps for pregnant women: Systematic search, evaluation, and analysis of features. *Journal of Medical Internet Research, 23*(10), Article e25667. https://doi.org/10.2196/25667

Friedman, B., & Hendry, D. G. (2019). *Value sensitive design: Shaping technology with moral imagination.* MIT Press.

Gendered Innovations in Science, Health & Medicine, Engineering, and Environment. (n.d.). *What is gendered innovations?* Stanford University. https://genderedinnovations.stanford.edu/what-is-gendered-innovations.html

Gqola, P. D. (2015). *Rape: A South African nightmare.* MFBooks Joburg.

Graham, S. S. (2022). *The doctor and the algorithm: Promise, peril, and the future of health AI.* Oxford University Press.

Grand View Research. (2023). *Pregnancy tracking and postpartum care apps market size, share, and trends analysis report by application (pre-partum, post-partum), by device (smartphones, tablets, others), by platform, by region, and segment forecasts, 2023–2030.* https://www.grandviewresearch.com/industry-analysis/pregnancy-tracking-post-partum-care-apps-market-report

Hannah-Jones, N., & The New York Times Magazine. (2021). *The 1619 project: A new origin story.* One World.

Hartman, S. (1997). *Scenes of subjection: Terror, slavery, and self-making in nineteenth-century America.* Oxford University Press.

HistoryoScience. (2021, January 25). *Londa Schiebinger, what is gendered innovations? Dec 30 2021* [Video]. YouTube. https://www.youtube.com/watch?v=2rAiVdj63_M&t=5s&ab_channel=HistoryoScience

Howell, F. M. (2023). A historical analysis of Black women's stratified reproduction and experiences of gendered racism in reproductive healthcare settings. *The Scholar & Feminist Online, 19*(2). https://sfonline.barnard.edu/a-historical-analysis-of-black-womens-stratified-reproduction-and-experiences-of-gendered-racism-in-reproductive-health-care-settings/

Hoyert, D. L. (2023). *Maternal mortality rates in the United States, 2021.* Centers for Disease Control and Prevention. https://www.cdc.gov/nchs/data/hestat/maternal-mortality/2021/maternal-mortality-rates-2021.htm

Jones, A., & Taylor, B. (2018). The role of mobile applications in pregnancy management: Evidence-based practice or commercial exploitation? *Healthcare Technology Review, 24*(2), 78–92. https://doi.org/10.1177/1077558720182412

Kaur, P., & Panneerselvam, D. (2023, July 24). *Bicornate uterus.* StatPearls. https://www.ncbi.nlm.nih.gov/books/NBK560859/

Kramer, K.-L. (2012). *User experience in the age of sustainability: A practitioner's blueprint*. Morgan Kaufmann.

Kumar, S., & Singh, P. (2020). Functionality versus usability: The pitfalls of feature creep in pregnancy apps. *Mobile Health Innovations Journal, 8*(4), 112–129. https://doi.org/10.1177/0987 654320200854

Lapperruque, J., Null, E., Crawford, A., & Brown, L. X. Z. (2022, June 24). *Following the overturning of* Roe v. Wade, *action is needed to protect health data*. Center for Democracy and Technology. https://cdt.org/insights/following-the-overturning-of-roe-v-wade-action-is-needed-to-protect-health-data/

Lau, P. L. (2024). AI gender biases in women's healthcare: Perspectives from the United Kingdom and the European Legal Space. In E. Gill-Pedro, & A. Moberg (Eds.), *YSEC yearbook of socio-economic constitutions* (Vol. 2023, pp. 247–274). Springer. https://doi.org/10.1007/16495_2023_63

Lazarevic, N., Pizzuti, C., Rosic, G., Boehm, C., Williams, K., & Caillaud, C. (2023). A mixed-methods study exploring women's perceptions and recommendations for a pregnancy app with monitoring tools. *npj Digital Medicine, 6*, Article 50.

Lewis, D. (2001). Representing African sexualities. In S. Tamale (Ed.), *African sexualities: A reader* (pp. 199–216). Pambazuka Press.

Lubin, K.-L. (2022). Conversations toward practiced AI—HCI heuristics. In J. Y. C. Chen, G. Fragomeni, H. Degen, & S. Ntoa (Eds.), *HCI International 2022—Late breaking papers: Interacting with eXtended reality and artificial intelligence* (Lecture Notes in Computer Science, Vol. 13518, pp. 377–390). Springer. https://doi.org/10.1007/978-3-031-21707-4_27

Lubin, K.-L. (2024, April 15). Wake up, humans! Our data crisis is really a humanity crisis—From bias to greed: The many data exploits. *Medium*. https://kemlaurin.medium.com/wake-up-humans-our-data-crisis-is-really-a-humanity-crisis-b260944ba20e

Lubin, K.-L., & Fan, L. T. (2025). Rethinking the rhetoric of surveillance in public safety: A critical discourse analysis. In K. Arai (Ed.), *Advances in information and communication: Proceedings of the 2025 Future of Information and Communication Conference* (Lecture Notes in Networks and Systems, Vol. 1285, pp. 657–677). Springer. https://doi.org/10.1007/978-3-031-84460-7_42

Lubin, K.-L., & Harris, R. A. (2024). Sex after technology: The rhetoric of health monitoring apps and the reversal of *Roe v. Wade*. *Rhetoric Society Quarterly, 54*(3), 247–262.

Lupton, D., & Pedersen, S. (2016). An Australian survey of women's use of pregnancy and parenting apps. *Women and Birth*, 29(4), 368–375. https://doi.org/10.1016/j.wombi.2016.01.008

McIlwain, C. D. (2019). *Black software: The internet and racial justice, from the AfroNet to Black Lives Matter*. Oxford University Press.

Mello, M. M., & Rose, S. (2024, March 7). Denial—Artificial intelligence tools and health insurance coverage decisions. *JAMA Health Forum*, 5(3), Article e240622. https://doi.org/10.1001/jamahealth forum.2024.0622

Milan, S., & Treré, E. (2020). The rise of the data poor: The COVID-19 pandemic seen from the margins. *Social Media + Society*, 6(3), 1–5. https://doi.org/10.1177/2056305120948233

Mole, B. (2023, November 16). *UnitedHealth uses AI model with 90% error rate to deny care, lawsuit alleges*. ArsTechnica. https://ars technica.com/health/2023/11/ai-with-90-error-rate-forces-elderly-out -of-rehab-nursing-homes-suit-claims/

Morgan, J. (2004). *Laboring women: Reproduction and gender in New World slavery*. University of Pennsylvania Press.

Nahra, K. J., Powell, B. A., & Dobkin, A. (2020, September 29). *California settles with Glow app over alleged privacy and security violations*. WilmerHale. https://www.wilmerhale.com/en/insights /blogs/wilmerhale-privacy-and-cybersecurity-law/20200929 -california-settles-with-glow-app-over-alleged-privacy-and-security -violations

Naim, N. (2023, July 28). *Women in the era of artificial intelligence: Increased targeting and growing challenges*. https://genderit.org /articles/women-era-artificial-intelligence-increased-targeting-and -growing-challenges

Natural Cycles. (2024, December 1). *Privacy policy*. https://www .naturalcycles.com/other/legal/privacy

Natural Cycles. (n.d.). *Birth control built around you*. https://www .naturalcycles.com/ca

Nissen, M., Huang, S.-Y., Jäger, K. M., Flaucher, M., Titzmann, A., Bleher, H., Pontones, C. A., Huebner, H., Danzberger, N., Fasching, P. A., Eskofier, B. M., & Leutheuser, H. (2024). Smartphone pregnancy apps: Systematic analysis of features, scientific guidance, commercialization, and user perception. *BMC Pregnancy and Childbirth*, 24, Article 782. https://doi.org/10.1186/s12884-024 -06959-1

Noble, S. U. (2018). *Algorithms of oppression: How search engines reinforce racism*. NYU Press.

Noble, S. (2021, October 29). *A conversation at the intersection of race, AI, and technology with Safiya Noble.* ISTAS21 Keynote Address. Moderated by L.-T. Fan, K.-L. Lubin, & J. S. Kim. University of Waterloo.

Obermeyer, Z., Powers, B., Vogeli, C., & Mullainathan, S. (2019). Dissecting racial bias in an algorithm used to manage the health of populations. *Science, 366*(6464), 447–453. https://doi.org/10.1126/science.aax2342

Pal, K. K., Piaget, K., Zahidi, S., & Baller, S. (2024, June 11). *Global gender gap report 2024.* World Economic Forum. https://www.weforum.org/publications/global-gender-gap-report-2024/digest/

Peña, P., & Varon, J. (2021, January 15). Descolonizando la inteligencia artificial: Un enfoque transfeminista [Decolonizing artificial intelligence: A transfeminist approach]. *Viento Sur.* https://vientosur.info/descolonizando-la-inteligencia-artificial-un-enfoque-transfeminista/

Perez, C. (2019). *Invisible women: Data bias in a world designed for men.* Abrams Press.

Piepgrass, S. C., Mirza, S., & Waltz, D. (2023, May 22). *AGs require company with ovulation tracking app to protect user data.* Regulatory Oversight. https://www.regulatoryoversight.com/2023/05/ags-require-company-with-ovulation-tracking-app-to-protect-user-data/

Presser, L., & Surana, K. (2024, November 25). A third woman died under Texas' abortion ban. Doctors are avoiding D&Cs and reaching for riskier miscarriage treatments. *ProPublica.* https://www.propublica.org/article/porsha-ngumezi-miscarriage-death-texas-abortion-ban

Roe v. Wade, 410 U.S. 113 (1973). https://www.oyez.org/cases/1971/70-18

Sherfinski, D. (2024, June 21). Data privacy concerns linger two years after end of *Roe v. Wade. Context.* https://www.context.news/money-power-people/data-privacy-concerns-linger-two-years-after-end-of-roe-v-wade

Society for Women's Health Research. (2025). *A timeline to making women's health mainstream.* https://swhr.org/about/1977-1989-timeline/

Spector-Bagdady, K., & Mello, M. M. (2022, June 30). Protecting the privacy of reproductive health information after the fall of *Roe v. Wade. JAMA Health Forum, 3*(6), Article e222656. https://jamanetwork.com/journals/jama-health-forum/fullarticle/2794032

Strand, N. H., Mariano, E. R., Goree, J. H., Narouze, S., Doshi, T. L., Freeman, J. A., & Pearson, A. C. S. (2021, June). Racism in pain medicine: We can and should do more. *Mayo Clinic Proceedings*, *96*(6), 1394–1400. https://www.mayoclinicproceedings.org/article/S0025-6196%2821%2900322-0/fulltext

Surana, K. (2024, September 16). Abortion bans have delayed emergency medical care. In Georgia, experts say this mother's death was preventable. *ProPublica*. https://www.propublica.org/article/georgia-abortion-ban-amber-thurman-death

Taylor, L., & Broeders, D. (2015, August). In the name of development: Power, profit and the datafication of the global South. *Geoforum*, *64*, 229–237. https://doi.org/10.1016/j.geoforum.2015.07.002

UNESCO. (n.d.). *Ethics of artificial intelligence: The recommendation.* https://www.unesco.org/en/artificial-intelligence/recommendation-ethics

United Nations. (2015). *Report of the International Conference on Population and Development, Cairo, 5–13 September 1994.* https://www.un.org/development/desa/pd/sites/www.un.org.development.desa.pd/files/icpd_en.pdf

U.S. Food and Drug Administration. (2025, March 6). *Human papillomavirus vaccine (HPV) safety.* https://www.cdc.gov/vaccine-safety/vaccines/hpv.html

Violence Against Women Act of 1994, H.R. 3355, 103rd Cong. (1994). https://www.congress.gov/bill/103rd-congress/house-bill/3355

Walter, C., & Tsang, A. (2022). *Patient data privacy at Clue: A statement from the co-CEOs.* Clue. https://helloclue.com/articles/about-clue/patient-data-privacy-at-clue-a-statement-from-the-co-ceos

Why data privacy is a concern in the wake of *Roe v. Wade* reversal. (2022, June 29). *CBS News.* ttps://www.cbsnews.com/news/abortion-ban-surveillance-tracking-technology/

Wixson, M. (2021, January 7). *Anti-racism and Dr. Susan Moore's legacy.* Michigan Medicine, University of Michigan. https://www.michiganmedicine.org/health-lab/anti-racism-and-dr-susan-moores-legacy

Women's Health Protection Act of 2023, H.R. 12, 118th Cong. (2023). https://www.congress.gov/bill/118th-congress/house-bill/12

Women's March. (2025). *Our vision.* https://www.womensmarch.com/about-us

Index

About the Author

Kem-Laurin Lubin is a Canadian-Caribbean design strategist, author, and thought leader specializing in sustainable technology design, artificial intelligence (AI) policy, and ethical innovation. Her work focuses on the intersection of AI, computational rhetoric, and human-centered design, with a particular emphasis on the societal and ethical implications of emerging technologies. She is recognized for advancing discourse around responsible AI and sustainable design, particularly in the context of marginalized and underrepresented communities.

Kem-Laurin's career spans nearly two decades across corporate leadership, academic research, and public advocacy. She has held prominent roles at Research in Motion (BlackBerry), Autodesk, and Siemens. At Autodesk, she led the design team behind Maya 3D Animation, a flagship tool in the entertainment and design industries, leading the product to its inaugural cloud instance. During her tenure at Research in Motion, she founded the company's first design research team and usability labs, driving early innovations in mobile technology and user experience.

In addition to her corporate achievements, Kem-Laurin is the author of *User Experience in the Age of Sustainability* (Kramer, 2012), a pioneering handbook that reframes design through the lens of ecological responsibility and sustainable innovation. The book has been adopted as a core text in undergraduate and graduate design programs worldwide.

Her academic work addresses the ways AI-powered systems categorize human behavior, as well as using limited data to reconstruct, represent, and characterize humans—a term she refers to as *algorithmic ethopoeia*. Her work highlights the ethical and social risks embedded in such technologies, and her other related publications include the following articles:

- "Conversations Towards Practiced AI—HCI Heuristics" (Lubin, 2022)
- "Sex After Technology: The Rhetoric of Health Monitoring Apps and the Reversal of *Roe v. Wade*" (Lubin & Harris, 2024)
- "Rethinking the Rhetoric of Surveillance in Public Safety: A Critical Discourse Analysis" (Lubin & Fan, 2025)

Kem-Laurin is also the founder of the AI Global South Summit and Human Tech Futures, initiatives dedicated to integrating socioeconomic realities and sustainability into AI design and governance. Her work focuses on amplifying the voices of the Global South—referred to by many leaders as the "Global Majority"—to ensure equitable technological development.

Kem-Laurin advises organizations on ethical AI practices and sustainable technology. She also contributes to shaping graduate curricula on design and AI ethics, mentoring the next generation of innovators. Her leadership continues to influence how technology is developed and deployed, promoting a future where design and AI are inclusive, sustainable, and socially responsible.

www.ingramcontent.com/pod-product-compliance
Lightning Source LLC
Chambersburg PA
CBHW060235030426
42335CB00014B/1466